A Practical Guide to Experimental Geometrical Optics

A concise, yet deep introduction to experimental geometrical optics, this book begins with the fundamental concepts and then develops the practical skills and research techniques routinely used in modern laboratories. Suitable for students, researchers, and optical engineers, this accessible text teaches readers how to build their own optical laboratory and to design and perform optical experiments. This hands-on approach fills a gap between theory-based textbooks and laboratory manuals, and allows the reader to develop their practical skills in this interdisciplinary field. It also explores the ways in which this knowledge can be applied to the design and production of commercial optical devices. Including supplementary online resources to help readers track and evaluate their experimental results, this text is the ideal companion for anyone with a practical interest in experimental geometrical optics.

YURIY A. GARBOVSKIY is Lab Manager of the UCCS BioFrontiers Center at the University of Colorado, Colorado Springs. His research interests include the physics of liquid crystals and nanomaterials. He is a Senior Member of the Optical Society of America.

ANATOLIY V. GLUSHCHENKO is Professor of Physics at the University of Colorado, Colorado Springs, and Director of the Center for Advanced Technologies and Optical Materials. He has received numerous awards in recognition of his research, including Inventor of the Year Award (University of Colorado, 2007) and the Thomas Jefferson Award (University of Colorado, 2013).

A Practical Guide to Experimental Geometrical Optics

Yuriy A. Garbovskiy

University of Colorado, Colorado Springs

Anatoliy V. Glushchenko

University of Colorado, Colorado Springs

CAMBRIDGE
UNIVERSITY PRESS

CAMBRIDGE
UNIVERSITY PRESS

University Printing House, Cambridge CB2 8BS, United Kingdom

One Liberty Plaza, 20th Floor, New York, NY 10006, USA

477 Williamstown Road, Port Melbourne, VIC 3207, Australia

314–321, 3rd Floor, Plot 3, Splendor Forum, Jasola District Centre, New Delhi – 110025, India

79 Anson Road, #06–04/06, Singapore 079906

Cambridge University Press is part of the University of Cambridge.

It furthers the University's mission by disseminating knowledge in the pursuit of education, learning, and research at the highest international levels of excellence.

www.cambridge.org
Information on this title: www.cambridge.org/9781107170940
DOI: 10.1017/9781316758465

First published 2017

Printed in the United Kingdom by TJ International Ltd. Padstow Cornwall

A catalog record for this publication is available from the British Library.

ISBN 978-1-107-17094-0 Hardback

Contents

Contents

Preface

Tell me and I forget. Teach me and I remember. Involve me and I learn.
Benjamin Franklin

The fields of optical science and engineering continue to thrive. Optical technologies have become a facet of modern society, penetrating our daily life to such an extent that even a layman, fascinated by the release of a new, colorful gadget, is convinced of the importance of optics. This same realization has occurred at the official level as well. The United Nations designated the year 2015 as The International Year of Light and Light-Based Technologies. This event made us think how we could contribute in honoring and disseminating the beauty of optics. After endless enjoyable discussions, both of us agreed that the most useful thing we could do was to provide the younger generation of optical experimentalists and novices in the field with a set of practical skills to assist in their exploration of optics.

Needless to say, this idea came to our minds a long time ago. Through our professional experience and teaching activities, we repeatedly faced the same problems while interacting with students. In short, even very knowledgeable students, who knew sign conventions perfectly and never made mistakes in their homework, suddenly became nervous in the lab as they struggled to identify basic optical components and create simple optical set-ups. Despite their theoretical knowledge, their lack of basic experimental skills made the students feel uncomfortable. We were curious why the same struggle faced all newcomers to experimental optics. Driven by our desire to find an answer to this question, we followed the universal scientific method and analyzed the existing literature. We read countless books on optics which, as we found out, all fall into two categories: theory-based textbooks and monographs, or lab manuals. It is generally believed that theory-based textbooks prepare students to later work in the lab, and any further guidance would come from lab manuals. A detailed analysis of several lab manuals which described optical experiments showed that, in many cases, the text assumes that the readers are familiar with basic concepts of experimental optics. Our daily interactions with students and with professionals from different fields, who are all trying to employ optical methods, revealed that this assumption is unfounded. In fact, it is quite natural that, when they begin, students lack the basic skills and understanding of how to work with basic optical elements. This lack of proper preparation results in lost time, wasted resources, and often leads to frustration.

From the pedagogical perspective, the aforementioned approach to teaching experimental optics treats students as *passive learners* because all of the included optical experiments are pre-designed. Therefore, the only thing the students can do is to take the described measurements. As a result, after completing such a "passive" lab course, most students lack any real comprehension of optical experiments because they never designed any experimental projects themselves! We decided to treat students as *active learners* by asking them to build their own optical lab and to design optical experiments. As we found out, this approach helps students develop important practical skills and makes them more confident during their work in the research labs. Moreover, it prepares students for the demands of the present-day job market, making them more competitive.

Inspired by the success of the students who went through our curriculum, we decided to write a book. The goal of this book is to fill the gap between the existing theory-based textbooks and the lab manuals by implementing the concept of active learning, with a focus placed on the development of experimental skills, which will then be transferable to any field of optical science and engineering. This goal is very ambitious, and in our attempt to achieve it, we plan to publish three books covering the experimental aspects of modern geometrical optics (Book 1), physical optics (Book 2), and their applications (Book 3).

This book deals with experimental geometrical optics only. Methodologically, the text is written so that it can easily be followed not only by readers with previous physics knowledge, but also by those who do not have a formal physics background. The logistics, manner, and depth of the material presented in this book are the results of our experience from working with many types of trainees in Europe, Asia, and the United States in our attempts to familiarize them with both classical and modern aspects of experimental geometrical optics.

Whether one is a college student, a researcher with no knowledge of optics, an engineer at a start-up, or an employee of a large corporation, this text will provide a rapid, yet deep, introduction to experimental optics. It has proven to be effective for the training of not only college and university students, but also those in the competitive workforce in general. This text will help readers realize that optics is an interdisciplinary field and to apply this knowledge to the production of new goods.

As to undergraduate and graduate students, who constitute the largest proportion of readers, this text will help them transition from a purely academic environment to one of industry by showing the connection between the academic disciplines of optics and the industrial applications of optical knowledge.

The first two chapters guide the reader through the market in modern optical components and industries. To start, they teach how to plan, design, budget, and set up optical experiments, which can be used by both faculty and students. Moreover, they lay out how to set up an optical laboratory, which is a necessary skill for practicing engineers and interns. Both the researcher and the engineer who have invested time to study websites and catalogs of optical companies (or in other words, the market in optical components and devices) are considered to be significantly more skilled than those who did not. Therefore, knowledge of the optical market is indispensable.

Since any optical experiment can be seen as a logical sequence of manipulations with light, such as producing, managing, detecting, and measuring this light, Chapters 3 and 4 first seek to make the reader familiar with various optical detectors. The text specifically addresses the dependence of the detected signals on the incident power/energy. Our experience shows that the users of this text, particularly undergraduate and graduate students, are thrilled that after reading these sections they understand how to build their own light detectors.

Chapters 5 through 8 pay great attention to the experimental characterization and quick, yet comprehensive, evaluations of simple optical components. In this particular case, these are lenses. The subsequent sections, in Chapters 9 and 10, deal with the use of these elements in the design of basic optical instruments, such as compound microscopes, telescopes, and different types of eyepieces. Further on, Chapter 11 is devoted to spherical mirrors, how to evaluate them, and their applications.

In Chapter 12, the reader will learn about optical aberrations and understand the limitations of the paraxial description of light propagation. Chapter 13 introduces elements of optical radiometry. In Chapters 14 and 15, the reader will sharpen their experimental skills by working with cylindrical lenses and measuring the refractive index of different materials. Chapter 16 provides an excellent overview of the knowledge acquired in the previous chapters by asking the reader to construct and study a prototype of a prism-based spectrometer.

Finally, in Chapter 17, the text concludes with an introduction to the world of modern optical software, such as TracePro, Zemax, OpTaliX, etc., which will broaden the readers' horizons and encourage them to use these modern tools in their future optical explorations.

The style and structure of each chapter is generic and dictated by the practical focus of the book. First, a list of the main objectives is provided.

Preface

The required "theoretical minimum" is covered in the background section. This section is followed by a list of materials and equipment needed to perform the suggested experiments. After that, a step-by-step procedure to accomplish the required experimental tasks is presented. To sharpen students' experimental and theoretical skills, chapters end with evaluation and review questions. In addition, they also offer experimental projects for further investigation and a list of recommended reading.

Acknowledgments

We are greatly indebted to our students and colleagues for their valuable comments and suggestions. Their positive feedback was a source of inspiration while working on this project. Finally, we would like to thank our families. Without their continuous support, patience, and encouragement, this project would never have been completed.

Markets for Optical Materials, Components, Accessories, Light Sources, and Detectors

1

Objectives

1. Analyze optical markets by studying web resources and specialized catalogs of leading optical companies.
2. Develop basic skills to search for optical components.
3. Estimate a budget for a simple optical set-up.

Background

Classical optical experimentalists made all their components and equipment (lenses, prisms, polarizers, mechanical holders, and even optical tables) by themselves. They spent many hours grinding and polishing a chunk of glass by hand to make a lens. As you can imagine, such an approach is time consuming and is not common today, but it brought substantial benefit to those early gurus of optics. *They were able to feel, deeply and internally, all aspects of their experiments.* Undoubtedly, this was one of the reasons why the classical texts on optics became classics! To familiarize you with those golden days of experimental optics, we recommend reading classical texts such as *Physical Optics* by Robert Wood, and *Procedures in Applied Optics* by John Strong. We intentionally placed these two references in the text here, instead of in the Further Reading section, to highlight the changes which experimental-based optics has made during the past few decades.

The emergence of optical companies completely changed the lives of optical scientists. Instead of polishing optical surfaces for weeks, they now bypass these hurdles by ordering required optical components from an optical company, leaving more time to concentrate on optical experiments. Knowing the optical markets and having experience in finding optical components needed is a very important skill for a modern optics researcher, not to mention that it is an important life skill in general.

As you may already know, optical companies have their own websites where information about their products is presented in a colorful way. Additionally, all optical companies print catalogs of their products which can be ordered free of charge. A researcher or optical engineer who has invested time, at least once, to study websites and catalogs of optical companies is considered to be significantly more skilled. Knowledge of the optical market is indispensable.

All optical companies can be broadly divided into two groups: Giant companies, which are the main players in the optical market and offer a large variety of optical equipment; and niche companies, which specialize in the production of specific equipment such as light detectors, optical phase retarders, and mechanical components. Optical companies have highly developed customer service departments. Researchers can send them questions online or call for assistance and customer service engineers will help.

The large number of optical companies and the variety of components they offer may at first appear daunting. After a certain investment of time, researchers develop an individualized method to search for necessary information and compare products from competing vendors. Special guides to optical companies also exist, such as search engines for optical components and companies. Among these guides, the "Photonics Corporate Guide" at www.photonics.com/BuyersGuide.aspx and the "Laser Focus World Buyer's Guide" at http://buyersguide.laser focusworld.com/ are very good resources. These web resources allow researchers to look for products and services (e.g. detectors, sensors, components, subassemblies, electronic and signal analysis instruments, fiber optics, accessories, imaging devices, cameras, displays, laser accessories, lasers, laser systems, light sources, etc.). Practically any optical company can be found using these links. In addition to the links we mention in the Further Reading section, there are a few more links to consider:

> www.coherent.com - Coherent (lasers)

> www.deltron.com - Del-Tron Precision Inc. (optical mechanics)

> www.pasco.com - PASCO (physics and engineering education)

> www.perkinelmer.com/home.aspx - Perkin Elmer (spectroscopy, light detection)

In addition to commercial websites, there are many professional social networks such as LinkedIn, Research Gate, and technical groups at

Facebook or Twitter offering great networking opportunities and valuable informational resources.

Materials Needed

(a) Materials and components for demonstration purposes:
 - Optical table, optical breadboard, or optical rail
 - Optical elements such as prisms, lenses, mirrors, polarizers, optical filters, diffraction gratings
 - Light sources such as tungsten filament lamps, mercury discharge lamps, He–Ne lasers, LEDs including those embedded in a smart-phone
 - Light detectors such as laser power meters, photoresistors, photodiodes, including features in a smart-phone e.g. CCD/CMOS camera
 - Mechanical components and holders
 - Screws and screwdrivers

(b) The tools necessary to perform the experiments in this particular chapter:
 - Catalogs of optical companies (Newport, Edmund Optics, Thorlabs, etc.)
 - Access to the Internet

Procedures

Exploring Web Resources

1. Go to the website of www.photonics.com ("Photonics Buyer's Guide") using the link shown here: www.photonics.com/ BuyersGuide.aspx.

2. Make yourself familiar with this web resource in the manner in which you would explore any other interesting webpage (for example, http://online.wsj.com/home-page), and figure out different options and ways to search for various optical components and equipment.

3. Try to find information on available tungsten filament lamps or white light sources. Use different ways to find such information: (i) Request information directly from the site of "Photonics Buyer's Guide" (to do this you will need to register); and (ii) explore the company's website and find information online.

4. Fill in Table 1.1.

> **Note:** The Evaluation and Review Questions section of this chapter is based on the information you should find while completing Tables 1.1–1.13. It is wise to review that section *before* you begin working with the tables and catalogs!

Table 1.1 Tungsten filament lamps

Item	Physical parameters[*]	Vendor	Catalog # & price
Tungsten filament lamp, type 1			
Tungsten filament lamp, type 2			
Tungsten filament lamp, type 3			

[*] Study specification data for the lamps and choose the parameters you think are important for the comparison.

Working with Catalogs

1. Make yourself familiar with the structure of each catalog provided. Find similarities and differences between them.

2. Find the following items/optical components and complete the tables below.

Table 1.2 Light sources (refer to Appendix 1A for an interesting comparison of emission spectra for some light sources)

Item	Vendor	Catalog # & price

Incandescent sources (white light):
    Tungsten filament lamp
    White LED (light emitting diode)		

Discharge lamps (UV light sources):
    Xenon lamp
    Mercury lamp		

Color LEDs (quasi-monochromatic light sources):
    Blue LED
    Red LED
    Green LED		

Lasers (monochromatic and coherent light sources):
    He–Ne (cw, or continuous wave laser at 632.8 nm)
    Nd:YAG (cw laser)
    Nd:YAG (pulsed laser)		

Table 1.3 Lenses

Item	Vendor	Catalog # & price
Double-convex lens		
Double-concave lens		

Table 1.3 (*cont.*)		
Item	**Vendor**	**Catalog # & price**
Plano-convex lens		
Plano-concave lens		
Cylindrical lens		

Table 1.4 Mirrors

Item	**Vendor**	**Catalog # & price**
Metallic mirror		
Dielectric mirror		

Table 1.5 Prisms

Item	**Vendor**	**Catalog # & price**

Table 1.6 Polarizers

Item	**Vendor**	**Catalog # & price**

Table 1.7 Beamsplitters

Item	**Vendor**	**Catalog # & price**

Table 1.8 Waveplates

Item	**Vendor**	**Catalog # & price**

Table 1.9 Diffraction gratings

Item	Vendor	Catalog # & price

Table 1.10 Filters

Item	Vendor	Catalog # & price
Band-pass filter		
Neutral density filter		
Interference filter		
Dichroic filter		

Table 1.11 Mechanics

Item	Vendor	Catalog # & price
Optical table		
Optical rail		
Breadboard		
Posts		
Post holders		
Lens holders		
Filter holders		
Translation stage		
Rotational stage		
Diaphragm		
Pin-holes		
Slits		

Table 1.12 Light detectors

Item	Vendor	Catalog # & price
Visible light detector (photodetector/photodiode)		
UV light detector		
IR light detector		
Photomultiplier		

Table 1.13 Accessories

Item	Vendor	Catalog # & price
Bolts and nuts (screws)		
Screwdrivers		
Optics cleaning tissues		
Gloves		
Alcohol solutions		

Evaluation and Review Questions

Estimate the budget necessary to build the following simple optical set-up, as shown in Fig. 1.1. Imagine you are a starting optical engineer at [your company name, Inc.] and you receive your first assignment, which is to test the reflection of a mirror for different angles of incidence and polarizations (parallel and perpendicular to the plane of incidence) at the wavelength of 632.8 nm. You realize that the company does not have any optical components/tables to perform the task and that the first thing you need to do is persuade management to allow you to purchase the necessary equipment. Write a half-page proposal on what should be purchased, why it is necessary, and how much it will cost.

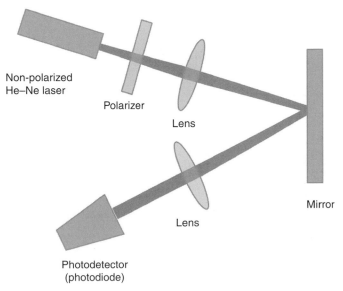

Figure 1.1 Example of a typical optical set-up.

Non-polarized
He–Ne laser

Polarizer

Lens

Mirror

Lens

Photodetector
(photodiode)

For Further Investigation

Assume that instead of the low-power (~500 mW/cm²) cw He–Ne visible ($\lambda = 632.8$ nm) laser shown in Fig. 1.1, you need to use a high-intensity (~80 MW/cm²) pulsed IR Nd:YAG laser ($\lambda = 1064$ nm). Which optical components will you need to check and/or replace in this case and why? Estimate the new budget.

Further Reading

Several companies provide an extensive array of optical components and

equipment. Some of these companies produce/sell general optical parts, such as lenses, prisms, holders, etc. Others specialize in a niche market, producing equipment or parts for specific applications. Here, we provide some examples of such companies:

www.edmundoptics.com - Edmund Optics
www.newport.com - Newport Corporation
www.thorlabs.com - Thorlabs

Appendix 1A

Figure 1A.1 Spectrum of an incandescent lamp.
Note: **Emitted power is shown in arbitrary units (a.u.); symbol 'λ' denotes wavelength of light.**

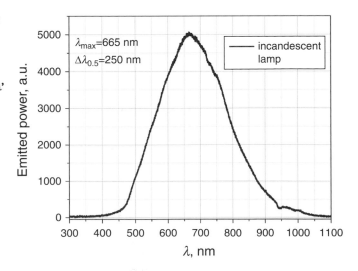

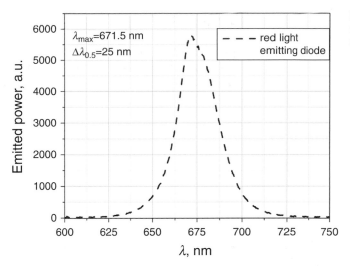

Figure 1A.2 Spectrum of a red light emitting diode.

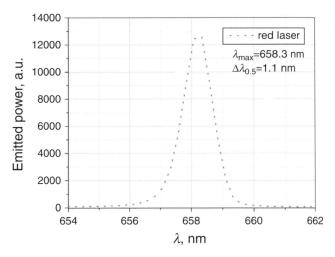

Figure 1A.3 Spectrum of a red laser.

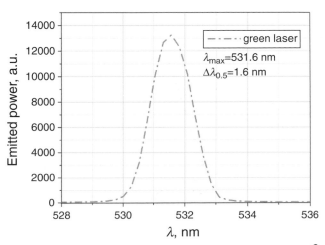

Figure 1A.4 Spectrum of a green laser.

**Figure 1A.5
Comparison of the
spectra of an
incandescent lamp, a
red light emitting
diode, a red laser, and
a green laser to
highlight the
differences in the
width of the
emissions.**

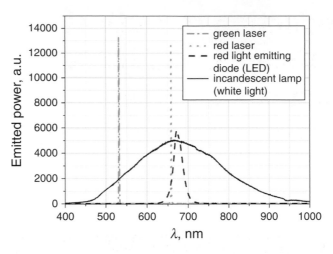

Introduction to Optical Experiments: Production, Management, Detection, and Measurement of Light

2

Objectives

1. Develop basic skills in handling optical components and assembling optical set-ups.
2. Develop skills in choosing light sources, light detectors, and components that manage light.
3. Using your favorite catalogs, "create" your own undergraduate optics laboratory.

Background

Any optical experiment can be seen as a logical sequence of manipulations with light. In general, experiments include basic steps such as producing, managing, detecting, and measuring light. Let's now consider each of these steps separately and in more detail.

Light Production

We use a variety of artificial light sources to create light. Natural sunlight, flames, and sparks were used in landmark optical experiments, including research that led to Nobel Prizes, even as recently as the twentieth century. Artificial light sources can be classified based on the physical mechanisms responsible for emitting the light.

(a) *Incandescent sources* Light is produced by a material heated to incandescence, and radiation arises from the de-excitation of the atoms or molecules in the material after they have been thermally excited. Black-body sources, tungsten-filament lamps, and tungsten–halogen lamps are all examples of incandescent light sources. Incandescent sources are widely used as white light sources.

(b) *Discharge lamps* Light is emitted during electrical discharge in a gas. Mercury and mercury–xenon lamps, xenon lamps, metal-halide lamps, high-pressure discharge lamps, low-pressure discharge lamps, long-discharge lamps, flash lamps, and fluorescent lamps are all examples. In addition to having narrow spectral lines, high-pressure discharge lamps have a continuous spectral background which is the result of greater molecular interactions. Low-pressure discharge lamps have sharper and narrower spectral lines and less intensive continuous spectral backgrounds compared to high-pressure lamps. Among the variety of discharge lamps, the so-called spectral lamps are of utmost importance. Examples of spectral lamps include mercury, zinc, cadmium, krypton, helium, thallium, and gallium spectral lamps. These lamps are characterized by narrow sharp spectral lines (for instance 0.05 Å spectral line width at 530 nm for a thallium lamp) and are widely used in spectroscopy and interferometry as calibrating lamps. Discharge lamps are commonly used as UV light sources.

(c) *Light emitting diodes (LEDs)* Light is emitted due to recombination of charge carriers in a semiconductor band gap (a semiconductor pair is needed for the light to be emitted).

(d) *Lasers* Light is emitted as a result of the stimulated emission of radiation by an excited ensemble of atoms. Some examples of lasers include solid-state, semiconductor, gas, liquid, plasma, and free-electron lasers. Lasers are unique sources of highly coherent and monochromatic light, and they are an indispensable tool in a variety of applications.

To choose an appropriate light source, the following basic parameters should be taken into consideration:

(a) *Spectral emission* The spectral emission can be monochromatic light (UV – ultraviolet lasers, visible (blue, green, red) lasers, IR – infrared lasers) or non-monochromatic light (UV light source, visible light source, IR light source). In experiments, there are no light sources which cover all wavelengths simultaneously from the UV to IR wavelength region. Theoretically, a black-body with extremely high temperature could be an ideal UV–visible–IR light source, but the required temperature is too high and exceeds the melting point of the electrodes and, therefore, it is impossible to construct such an ideal light source. For this reason, we select an appropriate light source with the required spectral output for each experiment. In many cases, we use discharge lamps for the UV region and incandescent sources for visible and IR regions. When we need a narrower wavelength distribution of the emission, laser and LED light sources are common choices.

(b) *Coherence* Light can be from either a coherent source, such as a laser, or an incoherent source. Incoherent light can be transformed to partially coherent light using additional optical equipment. For example, a simple set-up can be created using a light source, an interferometric filter, and a slit. The result of such a scheme is that, after passing through the filter and the slit, the light will be quasi-coherent.

(c) *State of polarization* There are randomly, partially, linearly, circularly, and elliptically polarized light sources. The state of polarization is easily changed by using polarizers and waveplates.

(d) *Output power/energy/intensity distribution (both spectral and spatial)* The term "energy" characterizes light sources operating in a pulsed regime. The term "power" or "intensity" is more appropriate for continuously operating light sources. Energetic-power output of a light source depends on the size of the source and has a non-uniform spectral distribution. High-powered light sources also require high-powered electrical supplies. The spatial distribution of the emitted light is described in terms of directionality of a light source.

(e) *Temporal regime* Light sources can be pulse or continuously (cw) operated depending on their design. By using special mechanical electro-optical modulators, one can transform continuous light sources into pulsed light sources. Intrinsically pulsed light sources are characterized by higher intensity compared to cw light sources. High-power cw light sources often require additional external cooling. Choosing between pulsed and cw light sources and their stability in time and space should be carefully considered.

(f) *Efficiency of the light output* Efficiency of light sources tells us how effectively external electrical energy is transformed into light energy. Such a transformation is spectrally non-uniform. Let us consider the following example:

- Total spectral efficiency of a typical flame-type carbon arc is ~60%.

- Spectral distribution and total efficiency of a carbon arc are: below 270 nm ~1%; 270–320 nm ~1%; 320–400 nm ~6%; 400–450 nm ~5%; 450–700 nm ~7%; 700–1125 nm ~10%; above 1125 nm ~30%.

Also, efficiency of light output can be considered in terms of luminous efficiency, measured in lm/W, where lm (lumen) is the SI unit of luminous flux and is a measure of the power of light perceived by the human eye.

Depending on a particular experimental task, an experimentalist must analyze available light sources and choose the most appropriate one. Once a particular light source characterized by the above-mentioned parameters is selected, one can modify the spectral emission, coherence, output intensity, and temporal regime, or even change them completely (state of polarization) by using special optical equipment. This will be discussed further in the next section, Light Management.

Table 2.1 shows estimated values for the basic parameters of commonly used light sources and can be used as a first guide to choosing an appropriate light source. More information about light sources and their properties can be found in specialized catalogs and on websites of optical companies (such as Newport, Edmund Optics, Thorlabs, etc.).

Light Management

During optical experiments, light is subjected to a variety of actions which can be called "light management." A light beam can be refracted, reflected, expanded, focused, collimated, split, delayed, scattered, attenuated, amplified, etc. To complete these actions, we use special optical devices composed of simpler optical components (lenses, mirrors, prisms, diffraction gratings, beamsplitters, optical filters, polarizers, waveplates, etc.). Each simple optical component performs one specific procedure with light, though sometimes with side-effects which need to be taken into consideration. By combining different components, an experimentalist can manage and direct light in a desired way. Table 2.2 briefly describes the main functions of optical components.

To choose the correct optical elements, first consider the following questions:

(1) *Spectral requirements* Is the optical element you are about to purchase transparent or suitable for its action in the studied optical interval?

Table 2.1 Basic parameters of commonly used light sources

	Spectral emission	Output power/energy/ intensity	Directionality and divergence	Efficiency (%) and luminous efficiency (lm/W)	Temporal regime	State of polarization	Coherence
Natural light sources							
Sunlight	0.2–2.6 μm	135.1 mW/cm^2 (solar constant)	isotropic in all directions		cw	randomly polarized	incoherent
Flame	0.4–2.6 μm*	common candle emits roughly 1 cd** (~1.464 mW/cm^2 at a distance of 1 cm)	isotropic in all directions		cw	randomly polarized	incoherent
Artificial light sources							
Tungsten filament lamp	0.3–3.0 μm	up to 2–3 kW (~40 cd/mm^2 or 2000–3000 mW/cm^2 at distance of 1 cm)	isotropic in all directions	up to 5% in visible region; ~15 lm/W	cw	randomly polarized	incoherent
High- and low-pressure mercury discharge lamp	0.3–0.75 μm	up to 5 kW	anisotropic, depending on lamp design (electrode shape, size, configuration, etc.)	up to 30–40% in UV; ~ 50 lm/W	cw and pulsed	randomly polarized	incoherent

Table 2.1 (*cont.*)

	Spectral emission	Output power/energy/ intensity	Directionality and divergence	Efficiency (%) and luminous efficiency (lm/W)	Temporal regime	State of polarization	Coherence
Fluorescent lamp	0.35–0.4 μm (black-light lamp); 0.525–0.70 μm (yellow lamp); 0.35–0.7 μm (white lamp)	up to 1–3 kW	isotropic	up to 22% in visible region; ~100 lm/W	cw	randomly polarized	incoherent
Light emitting diode	0.490 μm (blue) 0.66–0.69 μm (red) 0.56 μm green	~30–60 mW (average) and up to 1 W (high-power LEDs)	unidirectional	more than 40% in visible region; ~150 lm/W (average value)	cw	randomly polarized or polarized (depending on design)	incoherent
He–Ne laser	0.6328 μm	0.1–50 mW	beam divergence 0.5–2.5 mrad and beam diameter 0.5–2.5 mm	up to 0.1%	cw	polarized	coherent
Nd:YAG laser	1.064 μm	10–100 W (cw); up to 50 J (pulsed)	beam divergence 0.3–25 mrad and beam diameter 0.7–10 mm	0.1–2% (5–8%, diode pumped)	cw or pulsed (ns, ps, fs time duration)	polarized	coherent

* Depending on the temperature.

** The candela (cd) is the luminous intensity in a given direction of a source that emits monochromatic radiation of frequency 540×10^{12} hertz and that has a radiant intensity in that direction of $1/683$ watt per steradian (SI units).

Table 2.2 Simple optical components

	Action (what an element is for)	Types/composition	Wavelength region	Comments
Mirror	To reflect light	(a) Metallic (Al, Ag, Au) thin layer on a glass substrate	(a) 0.2–1.5 μm (Al) 0.325–1.5 μm (Ag) 0.65–1.5 μm (Au)	(a) Silver coating is the cheapest. Gold coating is commonly used in the IR region.
		(b) Dielectric multilayer films	(b) UV Visible IR	(b) Nearly 100% reflectivity at a particular wavelength
Prism	Dispersive element	LiF	0.11–7 μm	UV–visible–IR
	Light redirection	CaF_2	0.3–9 μm	near UV–visible–IR
		Glass	0.3–2.5 μm	near UV–visible–IR
		Quartz	0.185–2.7 μm	UV–visible–IR
		NaCl	2–15 μm	IR
		KCl	2–18 μm	IR
		KBr	5–25 μm	IR
Window	To transmit light	Windows and lenses can be of the same composition as prisms		
Lens	To transmit and refract light, causing light divergence or convergence			

Table 2.2 (cont.)

	Action (what an element is for)	Types/composition	Wavelength region	Comments
Diffraction grating	Dispersive element	Step gratings: (a) transmission (b) reflection Holographic gratings	Near UV–visible Far IR (up to 3000 μm)	
Beamsplitter	Partially reflecting and transmitting element	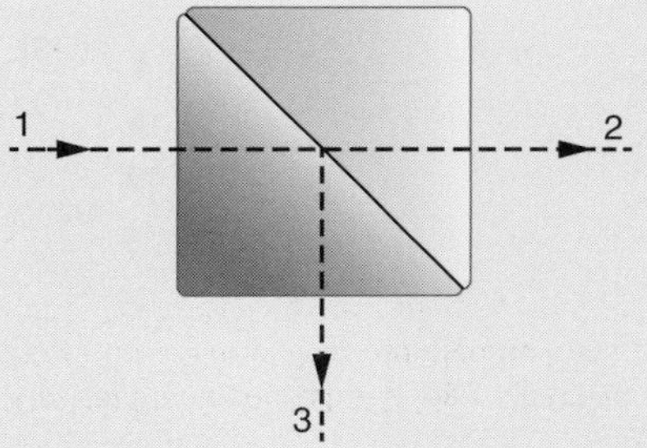	UV–visible–IR (depending on materials)	Beamsplitters can be divided into non-polarizing and polarizing. Polarizing beamsplitters reflect a large percentage of the S-polarized beam while transmitting the P-polarized beam.
Linear polarizer	To change state of polarization	(a) Absorptive polarizer (b) Beamsplitting polarizer (c) Birefringent polarizer (d) Thin-film (interferometric) polarizer	UV–visible–IR (depending on materials)	
Waveplate	To change optical phase shift	(a) Quarter-waveplate (b) Half-waveplate (c) Zero-order waveplate (d) Variable plate	UV–visible–IR (depending on materials)	Linear polarizer and waveplates can be combined to create and transform any state of polarization

18

Optical filter	To perform spectral filtering of light	Neutral density Absorptive Dichroic Monochromatic (interferometric)	UV–visible–IR (depending on materials)	Optical filters can be divided into long-pass, narrow band-pass etc.
Diffuser	To scatter light in order to make soft light	Ground glass, Teflon, holographic, opal glass, and grey glass	UV–visible–IR (depending on materials)	
Pin-hole	To transmit only a certain part (as a rule, the central part) of a light beam	Metal (steel), wood, plastic, black paper	UV–visible–IR	Material of pin-hole can be chosen depending on light intensity
Optical fiber	To transmit and carry light	Optical fibers consist of a high-refractive index core material surrounded by low-refractive index cladding material	UV–visible–IR	Heart of modern telecommunications

(2) *Intensity requirements* Can the light intensity you are using in your experiment damage any optical element? If the answer is yes, try to find an optical element with a much higher damage threshold. You should also consider the inverse situation when using extremely low light intensity. In this case, use so-called "blue optics," which are optical elements that are covered with antireflective layers to avoid losses caused by reflection.

(3) *Optical quality requirements* Is the optical element optically homogeneous enough, without aberrations, mechanical imperfections, etc.? Typically, optical quality of a surface is measured in a fraction of the wavelength of light (such as $\lambda/4$, $\lambda/10$).

(4) *Aperture requirements* Does the geometric size of the optical element fit the experimental goals?

(5) *Mounting requirements* What is the mounting requirement for an optical element? Do you need additional mechanical holders to do this?

Light Detection and Measurement

After being subjected to a sequence of actions and transformations, light should finally be detected and measured. To do this, a researcher must know all or some of the characteristics of the light, such as wavelength, energy (power, intensity), state of polarization, etc.

Devices which transform light energy into electrical signals (current or voltage) are called light detectors. Light detectors can be divided into thermal detectors (thermocouples and thermopiles, bolometers and thermistors, pyroelectric detectors, pneumatic or Golay) and quantum detectors (photoemissive detectors – photomultipliers, photoconductive detectors – photoresistors, semiconductor photodiodes, arrays of photodiodes – CCDs [charge-coupled devices]).

Each light detector can be specified by the following characteristic parameters:

- Type of photodetector

- What is measured (energy, power, intensity)

- Spectral selectivity (wavelength range of operation)

- Photodetector diameter (size of the photosensitive part of the light detector)

- Optical input (method by which the optical signal is delivered, such as a fiber cable [FC] or free space [FS])

- The maximum value of input signal, also called the damage threshold. The damage threshold is the maximum energy, power, or intensity which the detector can withstand.

- Noise equivalent power (NEP) must also be taken into consideration and is defined as NEP = $\Phi V_n / V_s$ where Φ is radiation flux, V_s is output signal, and V_n is root mean squared value of output noise.

- Detectivity (D) where $D = \sqrt{A \cdot \Delta f}/$NEP, where A is the area of the detector and Δf is the noise-equivalent frequency bandwidth. NEP and detectivity determine a detection threshold which is the minimum power of the detected optical signal.

When choosing appropriate light detectors, all the above-mentioned parameters should be considered. Since the light detector is connected to a digital/analog voltmeter, computer, or other external device, review the displayable values of the measuring device.

Great attention should be paid to the electrical output and output impedance of the light detectors. Connections should be made using special connecting cables. Commonly used light detectors and their basic parameters are shown in Table 2.3.

Using the described light detectors, the energetic parameters (energy, power, intensity) of light can be measured. Spectrometers are used to measure the wavelength distribution of the intensity of the detected light, or light spectra. These devices are described in the subsequent chapters of the text.

Mechanics of an Optical Experiment

Thus far, we have discussed light sources and basic optical components in a "theoretical" manner. We have been omitting such crucial practical questions as "how should we assemble all these optical components into a set-up?" A snapshot of any optical experiment demonstrates the following common picture: Light sources, light detectors, and optical elements are mounted using special mechanical holders placed on an optical table. Therefore, to make an optical experiment, experimentalists should have in their possession an appropriate light source, a light detector suited to the light source, optical elements, mechanical holders to mount the optical elements, and an optical table to position all the above-mentioned optical and mechanical components.

An optical table (or a rail, at least) is very necessary for any optical experiment. Since the beginning of optics, Herculean efforts have been made in the field of optical table design, and these efforts continue today. Modern optical tables have complicated structures and are expensive (~US$15,000 for one unit). The general goal of such complexity in an

Table 2.3 Light detectors

	Spectral sensitivity	Response time	D (cm Hz$^{1/2}$ W^{-1})	Optical input or linear range	External power supply requirements
Thermal photodetectors					
Thermocouples and thermopiles	0.8–40 µm	10–30 ms	10^8–10^9	10^{-10}–10^{-8} W	yes
Thermistor (bolometer)	0.8–40 µm	10–30 ms	10^8–10^9	10^{-6}–10^{-1} W	yes
Pyroelectric detectors	0.8–1000 µm	5–1000 ms	10^7–10^8	10^{-6}–10^{-1} W	yes
Pneumatic or Golay cell	0.8–1000 µm	2–50 ms	10^8–10^9	10^{-6}–10^{-1} W	yes
Quantum photodetectors					
Photomultipliers	0.2–1.0 µm	0.3–15 ns	10^{12}–10^{18}	5.0–6.0 (decades)	yes
Photoresistors	0.75–6.0 µm	50 ns – 1 ms	10^9–10^{12}	5.0–6.0 (decades)	yes
Photodiodes	0.4–5.0 µm	1 µs – 1 ms	10^8–10^{12}	3.0–4.0 (decades)	yes
Arrays of quantum photodetectors					
CCD cameras	Near UV – visible – near IR	Typically, the response time of CCD and CMOS is determined in frames/ second (fps)	All these parameters depend on the exact camera and applications the camera is designed for		yes
CMOS cameras					

optical table's structure is to minimize vibrations and provide stability for the optical elements (acoustic and mechanical vibrations, air movement, and even breathing [do not chat much during an experiment as it can affect your results dramatically!]). All types of optical tables have two common features: they are massive and very mechanically stable. Optical companies offer a variety of optical tables, breadboards, vibration-isolation workstations, rails, and different types of mechanical holders (low-cost post holder, cam-lock post holder, stainless-steel posts, post clamps, etc.). Detailed descriptions of these can be found in specialized catalogs.

Optical elements may be assembled on an optical table in a few different ways. The most common assembly is either (1) mechanical holders with optical elements are fixed directly onto an optical table (surface of optical tables consists of arrays of holes, and any mechanical holder can be attached to a hole using a screw); or (2) mechanical holders with optical elements are fixed on an optical rail, which is then attached to the surface of an optical table.

How to Handle Optical Elements

Optical elements are incredibly delicate and sensitive to external contaminants such as dust, powders, liquid stains, etc. Keep in mind that optical elements are too expensive to be handled inappropriately. Often, dirty optics cannot demonstrate appropriate performance requirements, which affects an experiment dramatically. Always try to avoid contaminating optical elements. Never touch them using your fingers. Natural chemical compounds such as fat and water contaminants can completely ruin an optical element. One should use gloves or finger-cots to place optical elements into mechanical holders. After your experiment is complete, try to keep optical elements in isolated boxes. Also remember the golden rule of optics – if it's not dirty, don't clean it.

Creating an Optical Experiment

Your skill in performing an optical experiment can gradually improve only by conscious, continuous, and hard experimental work. Sitting and thinking instead of practically doing will not help. Again, remember that serious optical experimentalists develop and improve their skills not only by manipulating commercially available optical elements but by fabricating optical components they need by themselves... ☺

Materials Needed

- Optical table or optical rail
- Optical elements: Prism, lens, mirror, polarizer, optical filter, diffraction grating
- Light sources: Tungsten filament lamp, He–Ne laser, LED, etc.
- Light detectors: Laser power meter, photoresistor, photodiode
- Mechanical holders
- Screws and screwdrivers

Procedure

1. Identify all components of the optical experiment: Light sources, light detectors, available optical components and mechanical holders, optical table and optical rails.

2. Complete the following tables: Include only the equipment that is available in the lab.

Table 2.4 Light sources

Light source	Spectral emission	Output power/ energy/ intensity	Directionality and divergence	Efficiency (%) and luminous efficiency (lm/W)	Temporal regime	State of polarization	Coherence

Table 2.5 Optical elements

Optical element	What the element does to light	Wavelength region	Comments

Evaluation and Review Questions

Refer to the available optical company catalogs. Assume that you need to develop a nice optical laboratory for undergraduates. Estimate the budget necessary to create one station in the lab using the catalogs.

Table 2.6 Components for an optical lab

Item	Quantity	Price	Company
Optical table			
Breadboard			
Optical rail			
Mechanical holders			
Accessories			

Table 2.6 (*cont.*)

Item	Quantity	Price	Company
Optical elements			
Light sources			
Light detectors			

For Further Investigation

1. Propose several variants of the budget: expensive, moderate, and cheap. Your final results should be presented in the form of a table.

Table 2.7 Budgeting an optical lab

Budget	Total price	Available facilities	Equipment provider
Expensive			
Moderate			
Cheap			

2. Begin your own optical glossary. Write definitions for the optical terms you studied in this chapter.

Further Reading

General

E. Hecht, *Optics*, 4th edition, New York: Pearson/Addison Wesley, 2001

F. L. Pedrotti, S. J. L. Pedrotti, L. M. Pedrotti, *Introduction to Optics*, 3rd edition, Upper Saddle River, NJ: Pearson Prentice Hall, 2007

Specialized

Handbook of Optics, W. G. Driscoll (editor), W. Vaughan (associate editor), New York: McGraw-Hill, 1978

Springer Handbook of Materials Measurement Methods, H. Czichos, T. Saito, L. Smith (editors), Berlin: Springer, 2006

G. Rieke, *Detection of Light from the Ultraviolet to the Submillimeter*, 2nd edition, New York: Cambridge University Press, 2003

Catalogs

www.coherent.com/ - Coherent
www.deltron.com/ - Del-Tron Precision Inc.
www.edmundoptics.com/ - Edmund Optics
www.newport.com/ - Newport Corporation
www.perkinelmer.com/home.aspx - Perkin Elmer
www.thorlabs.com/ - Thorlabs

Appendix 2A

Table 2A.1 Basic radiometric and photometric units

| | Spatial density | | | |
Basic concept	Areal density at a surface	Intensity	Specific intensity	Volumetric density
Radiometric units				
Radiant energy, joules, J				Radiant density, joules per cubic meter, J/m³
Radiant flux, watts, W	Radiant exitance, irradiance, watts per square meter, W/m²	Radiant intensity, watts per steradian, W/sr	Radiance, watts per steradian-square meter, W/(sr m²)	

Table 2A.1 (*cont.*)

Basic concept	Spatial density: Areal density at a surface	Spatial density: Intensity	Spatial density: Specific intensity	Spatial density: Volumetric density
Photometric units				
Luminous energy, lumen seconds [talbot], lm s				Luminous density, lumen seconds per cubic meter, lm s/m^3
Luminous flux, lumens, lm	Luminous exitance, illuminance, lux (lx) lumens per square meter, lm/m^2	Luminous intensity, candela, cd [lumens per steradian], lm/sr	Luminance [photometric brightness], lm/(sr m^2) = cd/m^2	
Definitions				
Power emitted by a light source into the whole space per time unit	Power per detector area	Power emitted by a light source into a solid angle	Power emitted by the Lambert irradiator (uniformly diffusing surface, for which the luminance is the same in all directions) of a surface into a solid angle	Energy per unit volume

Unit Conversions

Radiant flux	1 W (watt) = 683.0 lm at 555 nm
	1 J (joule) = 1 W s (watt second)
Luminous flux	1 lm (lumen) = 1.464×10^{-3} W at 555 nm = $1/(4\pi)$ cd (candela) (only if isotropic)
	1 lm s (lumen second) = 1 T (talbot) = 1.464×10^{-3} J at 555 nm

3 Light Detectors Based on Semiconductors: Photoresistors, Photodiodes in a Photo-Galvanic Regime – Principles of Operation and Measurement

Objectives

1. Study the types, structure, and operational principles of simple light detectors (photoresistors, photodiodes working in a photo-galvanic regime).
2. Design an electrical circuit with a photoresistor and measure the dependence of photocurrent on input light power.
3. Design an electrical circuit with a photodiode operating in photovoltaic mode. Plot the dependence of photovoltage on input light power.
4. Develop practical skills in making light detectors based on semiconductors.
5. Estimate the budget of the considered semiconductor light detectors for an undergraduate optics lab.

Background

As was shown in Chapter 2, there are different types of light detectors: devices measuring light energy and/or power. In this and the following chapter, we consider light detectors based on semiconductors, which are very popular among researchers. In fact, semiconductor light detectors are the necessities of modern optical research life.

Semiconductor light detectors (photoconductors or photoresistors, and photodiodes) are typical quantum light sensors in which light beam energy is converted into free charge carriers generated through an internal photo-effect. The current and/or voltage due to these carriers can be measured directly. Compared to other quantum detectors, such as photomultipliers, photoresistors and photodiodes have low geometrical size and cost, are sensitive in the near UV–visible–near IR region, do not require high-voltage power supplies and, as a result, are widely used in modern research. A common (but not critical) disadvantage of all semiconductor detectors is larger noise compared to photomultipliers. If you need to detect extremely low light intensity ("photon counting"), you should use a photomultiplier instead of a semiconductor light detector. However, in real routine optical research, "photon counting" is not a common procedure (unless your research is in the field of extremely low electromagnetic fields). Semiconductor light detectors are much more practical and demonstrate excellent performance.

The measurable quantity produced by a semiconductor light detector is electric current or voltage. The dependence of such a measured electrical signal upon light power can be linear or non-linear. These two different regimes of light detector operation, namely the linear and non-linear dependence of the measured quantity (current or voltage) on the light power, will be discussed using practical examples of photoresistors and photodiodes.

Photoresistors

Semiconductors represent a broad class of materials which have electrical resistivity in the range between those of typical metals and typical insulators (i.e. between about 10^{-3} and 10^{-9} Ω cm). The physical properties of semiconductors can be described by the quantum theory of the electronic energy band structure of solids. Omitting an exhaustive treatment of this theory by referring readers to the specialized literature, we emphasize only the most important theoretical conclusions. According to them, in semiconductors we deal with conduction and valence energy bands – two bundles of energy levels – separated by a forbidden region, a fundamental energy gap (E_g) (Fig. 3.1).

The conduction band has a higher energy than the valence band. Electrons at the valence band are bonded and cannot move; thus, no current flows through the material. Electrons at the conduction band are "free," and when a small voltage is applied, they move, constituting current. In other words, *to induce material to conduct current, one needs to populate the conduction band with electrons*. However, one obstacle

Figure 3.1 Typical diagram of the electronic energy band structure of solids.

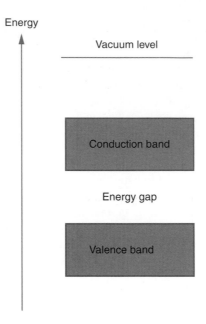

stands in the way: The energy gap. The value of E_g determines the conductive (resistive) properties of the material. Good conductors have no gap between the valence and conduction bands, good insulators have a big energy gap, and semiconductors have a gap somewhere in between. Indeed, diamond, a good insulator, has an energy gap of around 6 eV (electronvolt), while silicon (Si) and germanium (Ge), the most popular semiconductors, have gaps of 1.17 eV and 0.775 eV, respectively.

A photoresistor is a semiconductive material whose resistance decreases with increasing incident light intensity. The resistance of a photoresistor non-illuminated with light is called the *dark resistance*. The resistance of a photoresistor illuminated with light is called the *photo-resistance*. Light absorbed by a photoresistor generates electron–hole pairs. As a result, the concentration of the free charge carriers increases and resistance decreases. Photoresistors absorb light only in a certain wavelength diapason. To be absorbed, the energy of a light photon should not be smaller than the energy gap. The most common photoresistor materials and spectral diapason of their sensitivity are shown in Table 3.1.

For use as a light detector, a photoresistor can be connected in series with a voltage supply, a variable resistor, and a microammeter. It should be noted that the ammeter can be replaced with a reference resistor and a voltmeter connected in parallel. Figure 3.2 shows a general scheme for the electrical circuit of a light detector based on a photoresistor.

Table 3.1

Photoresistor material	Spectral diapason (µm)
Cadmium sulfide (CdS)	0.4–0.8
Cadmium selenide (CdSe)	0.5–1.0
Lead sulfide (PbS)	0.8–3
Lead selenide (PbSe)	1–5

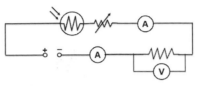

Figure 3.2 General scheme for the electrical circuit of a light detector based on a photoresistor.

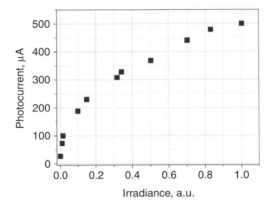

Figure 3.3 Photocurrent through photoresistor vs. irradiance. The irradiance is controlled by means of neutral filters and expressed using arbitrary units (a.u.).

Initially, when a photoresistor is not illuminated with light, so-called *dark current* will flow through the electrical circuit. As a result of light illumination, the electrical current through a circuit increases due to decreasing resistance of the photoresistor. *Photocurrent* I_{ph} can be defined as the difference between current I through an electrical circuit during light illumination and the dark current I_{dark}: $I_{ph} = I - I_{dark}$ (it can be measured with a microammeter). A typical dependence of photocurrent on irradiance is shown in Fig. 3.3.

As seen in Fig. 3.3, the experimental dependence is very non-linear, and for this reason, such a regime of light detection can be termed non-linear (please do not confuse this term with anything related to non-linear optics – this is a whole different fun).

Figure 3.4 Typical structure of photodiodes. Left: Conventional design (p–n junction). Right: Improved p–i–n structure provides faster response time (i denotes a thin layer of insulator sandwiched between p and n regions of the photodiode).

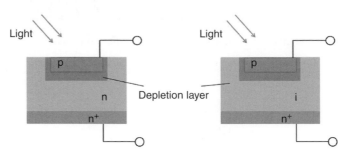

Photodiodes

A photodiode is a p–n junction or p–i–n structure (Fig. 3.4). A p–n junction is formed by joining p-type and n-type semiconductors in very close contact. When a photon of sufficient energy strikes the diode, it excites an electron, thereby creating a *free electron* and a *hole*. If the absorption occurs in the junction's depletion region, or one diffusion length away from it, these carriers are swept from the junction by the built-in field of the depletion region. Thus, holes move toward the anode, and electrons toward the cathode, and a photocurrent is produced.

Typical photodiode materials and the spectral diapason of their sensitivity are shown in Table 3.2.

As light detectors, photodiodes can operate in so-called *photovoltaic* and *photocurrent* modes. The photovoltaic mode does not require an external source of bias voltage while the photocurrent mode needs a reverse bias. In the photovoltaic mode, incident light generates a voltage across the terminals of the photodiode. The photovoltaic mode is a result of the internal photovoltaic effect, which is the basis of solar cells. In photovoltaic mode, the response curve of a photodiode (light-induced voltage) is very non-linear (Fig. 3.5).

Table 3.2

Photodiode material	Spectral diapason (μm)
Silicon (Si)	0.19–1.1
Germanium (Ge)	0.4–1.7
Indium gallium arsenide	0.8–2.6

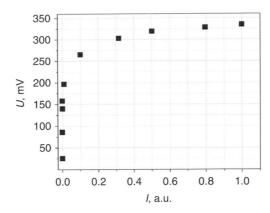

Figure 3.5
Photodiode SD-444-12 in photovoltaic regime (light source: green 35 mW cw laser). Intensity of the laser beam is controlled by means of neutral filters and expressed in arbitrary units (a.u.).

Materials Needed

- Optical table/optical rail
- Optical attenuators (neutral density filters). They are labeled #1, #2, #3, #4.
- Light sources (tungsten filament lamp, red cw laser, LED)
- Photoresistor, such as PDV-P8001 from Advanced Photonics
- Photodiode, such as BPW 34 from Opto Semiconductors
- Proto-board
- Resistance capacitance box

- Two-channel DC voltage supply
- Electrical cables and wires
- Multimeter
- Mechanical holders
- Screws and screwdrivers
- Catalogs of optical companies (Newport, Edmund Optics, Thorlabs, etc.)
- Specification sheets for semiconductor light detectors used (photoresistors and photodiodes)

Procedures

Measuring Light Using a Photoresistor

1. Read the specification sheet for cadmium sulfide photoresistors (see Further Reading). Complete the following table:

Table 3.3

Available CdS photoresistors	Interval of maximum voltage (V)	Operating temperature (°C)	Spectral peak (nm)	Interval of dark resistance (MΩ)	Response time (ms) Rise	Response time (ms) Decay	Maximal frequency (Hz)

Maximal frequency f_{max} can be estimated using the following expression: $f_{max} = $ (rise time + decay time)$^{-1}$.

(a)

(b)

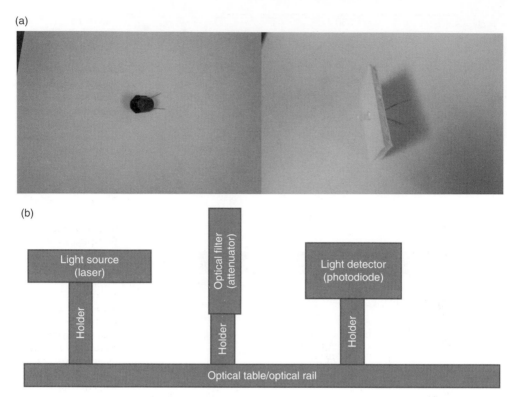

Figure 3.6
(a) A photoresistor (left) and an LED (right). (b) Optical set-up to characterize a photoresistor light detector.

2. Identify a photoresistor available in your optics kit and a suitable light source (either a laser/laser pointer or an LED) (Fig. 3.6a).

3. Assemble the optical set-up shown schematically in Fig. 3.6b. The light detector (photoresistor) should be connected to a power supply, as shown in Fig. 3.7a. If using a laser pointer as the light source, simply put it in a holder. If using an LED as the light source, use the connection scheme shown in Fig. 3.7b.

4. Block the light-sensitive face of the photoresistor, wait for ~30–60 s. Choose the resistance on the 100 kΩ resistor box. Only after that, apply DC 20 V. By changing the resistance of the resistor box, estimate the dark current through the photoresistor (Fig. 3.7c or 3.7d). Complete Table 3.4. (It is important to keep the light-sensitive face of the photoresistor blocked during measurements.)

Table 3.4

Applied DC (V)	Resistance in series (Ω)	Dark current (μA)	Dark resistance* (MΩ)
20	100 000		
20	1 000		

Table 3.4 *(cont.)*			
Applied DC (V)	Resistance in series (Ω)	Dark current (μA)	Dark resistance[*] (MΩ)
20	100		
20	15		
			Average dark resistance

[*] $R_{dark} = (U_{applied}/I_{dark}) - R_{in\ series}$, where R_{dark} is dark resistance, $U_{applied}$ is applied DC voltage, I_{dark} is dark current, $R_{in\ series}$ is resistance in series.

(a)

(b)

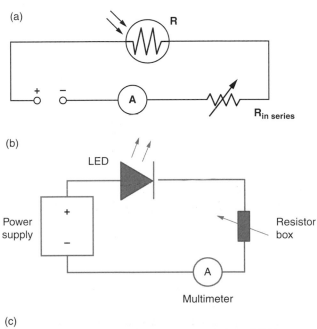

(c)

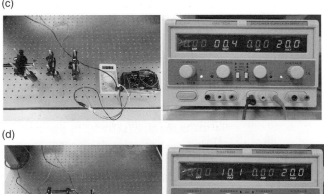

(d)

Figure 3.7 (a) Electrical circuit of light detector based on photoresistor. **(b)** Schematic of a light source based on an LED (Typical parameters: $U = 10\,V$, $R = \sim10\,k\Omega$. Do not exceed current through the LED of 50mA; the amount of current is controlled using a resistor box and is measured with a multimeter. **(c)** Estimating the dark current through the photoresistor; laser pointer is used as a light source. **(d)** Characterizing a photoresistor as a light detector using an LED as a light source.

Depending on the photoresistor model, you may discover that dark photocurrent is not detected by your apparatus (for example, multimeter). This means that the real photocurrent is below the limit your apparatus can measure. In this case, do not be frustrated. It is still possible to estimate the minimal value of dark resistance of the photoresistor. How do you do this? Let us assume that your apparatus can measure a minimal current of 1 μA (as a rule, all types of available multimeter can do this easily). Since dark photocurrent is not detected, this indicates that the real photocurrent is smaller than 1 μA and, according to the expression below Table 3.4, dark resistance is larger than $(U_{applied}/10^{-6}\,A) - R_{in\ series}$. Finally, add to Table 3.4, "Dark resistance $> \ldots$ MΩ."

5. Uncover the light-sensitive face of the photoresistor and measure the photocurrent due to daylight. Complete the following table.

Table 3.5

Applied DC (V)	Resistance in series (Ω)	Photocurrent (μA)	Photoresistance* (kΩ)
5	100		
5	10 000		
10	100		
10	10 000		
15	100		
15	10 000		
20	100		
20	10 000		

* $R_{photo} = (U_{applied}/I_{photo}) - R_{in\ series}$, where R_{photo} is photoresistance, $U_{applied}$ is applied DC voltage, I_{photo} is photocurrent, $R_{in\ series}$ is resistance in series.

Compare the obtained values of photoresistance and dark resistance of the photoresistor.

Switch on the laser/LED and, by placing different optical attenuators (#1, 2, 3, 4), measure the dependence of photocurrent on irradiated light power. Complete Table 3.6.

Transmittance of the attenuators for a selected laser/LED wavelength (~630–660 nm for red light source, and ~530–560 nm for green light source) can be found from transmittance spectral graphs.

Table 3.6

Applied DC (V)	Resistance in series (Ω)	Attenuator	Transmittance (%)	Power (a.u.)	Photocurrent (daylight) (μA)	Photocurrent (laser/LED light + daylight) (μA)	Photocurrent (laser/LED light) (μA)	Photoresistance (kΩ)
10	500	NO	100	1				
		#1						
		#2						
		#3						
		#4						
20	500	NO	100	1				
		#1						
		#2						
		#3						
		#4						

6. Plot the dependence of photocurrent on light power.

7. Plot the dependence of photoresistance on light power.

Measuring Light Using Photodiodes

Photovoltaic mode

1. Read the specification sheet for the photodiode BPW 34 (see Further Reading). Complete the following table.

Table 3.7

Photodiode area (mm x mm)	Interval of spectral sensitivity (nm)	Spectral peak (nm)	Spectral sensitivity (A/W)	Quantum yield	Ambient temperature (°C/°F)	Dark current (pA)	Capacitance (pF)	Response time (μs)		Maximal frequency (Hz)
								Rise	Decay	

Figure 3.8 Electrical circuit of a light detector based on a photodiode: Photovoltaic mode of operation.

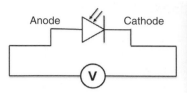

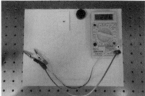

2. Using the multimeter, identify the cathode (−) and anode (+) of the photodiode and design the electrical circuit shown in Fig. 3.8. To do this, illuminate the photodiode with any ambient light source. If the multimeter shows a positive value, the wire connected to a common terminal of the multimeter is the cathode of the photodiode.

3. Assemble the optical set-up shown schematically in Fig. 3.6b (the photoresistor should be replaced with a photodiode connected to a voltmeter).

4. Open the light-sensitive face of the photodiode and measure the voltage across the photodiode generated by daylight.

5. Switch on the laser/LED and, by placing different optical attenuators (#1, 2, 3, 4), measure the dependence of the voltage across the photodiode on irradiated light power. Complete the following table.

Table 3.8

Attenuator	Transmittance (%)	Power (a.u.)	Voltage across photodiode (daylight) (mV)	Voltage across photodiode (laser/LED light + daylight) (mV)	Voltage across photodiode (laser/LED light) (daylight) (mV)
NO	100	1			N/A
#1					N/A
#2					N/A
#3					N/A
#4					N/A

6. Plot the dependence of measured voltage on light power.

Evaluation and Review Questions

1. Is the photoresistor sensitive to the polarity of the applied voltage? Why?
2. Is it a good idea to block the light-sensitive face of the photoresistor using your fingers? Consider the possible consequences of such an activity. How could it influence the experimental results?
3. When you irradiate the photodiode operating in photovoltaic mode, why do you observe saturation of the measured photovoltage?

Further Reading

General

F. L. Pedrotti, S. J. L. Pedrotti, L. M. Pedrotti, *Introduction to Optics*, 3rd edition, Upper Saddle River, NJ: Pearson Prentice Hall, 2007

Specialized

Handbook of Optics, W. G. Driscoll (editor), W. Vaughan (associate editor), New York: McGraw-Hill, 1978

K. F. Brennan, *The Physics of Semiconductors with Applications to Optoelectronic Devices*, New York: Cambridge University Press, 2003

L. Desmarais, *Applied Electro-Optics*, Upper Saddle River, NJ: Prentice Hall, 1997

G. Rieke, *Detection of Light from the Ultraviolet to the Submillimeter*, 2nd edition, New York: Cambridge University Press, 2003

B. G. Yacobi, *Semiconductor Materials: An Introduction to Basic Principles*, New York: Kluwer Academic/Plenum Publishers, 2003

Chapter 2 of this book

Useful Links

PDV-P8001 Datasheet (HTML) from Advanced Photonix, Inc.: www.alldatasheet.com/datasheet-pdf/pdf/237784/ADVANCEDPHOTONIX/PDV-P8001.html

Silicon PIN photodiode with enhanced blue sensitivity, in SMT Datasheet: www.osram-os.com/Graphics/XPic2/00131269_0.pdf

4 Linear Light Detectors Based on Photodiodes

Objectives

1. Study the types, structure, and operational principles of linear semiconductor light detectors.
2. Design an electrical circuit and test a photodiode as a light detector operating in photoconductive (reverse biased) mode. Measure the dependence of photocurrent on incident (input) light power, and find the dynamical diapason of such a light detector.
3. Design an electrical circuit and test a photodiode as a light detector operating in photoconductive (reverse biased) mode with an external operational amplifier. Measure the dependence of detected voltage on incident (input) light power for different feedback resistors.
4. Using a linear light detector, verify the "inverse square law" by measuring the irradiance of the photodiode vs. the distance between the incandescent light source and the light detector.
5. Develop practical skills to make convenient and reliable linear light detectors based on semiconductors.
6. Estimate the budget of linear semiconductor light detectors for an undergraduate optics lab.

Background

Photoconductive Mode of Photodiodes

In photoconductive mode, a diode is often reverse biased, which dramatically reduces the response time at the expense of increased noise. This increases the width of the depletion layer, which decreases the junction's capacitance, resulting in faster response times. The reverse bias induces only a small amount of current (known as saturation or back current) along its direction while the photocurrent remains virtually the same. The photocurrent is linearly proportional to the irradiance (Fig. 4.1), thereby the light detector is linear.

The photoconductive mode of a diode can be used both without an external amplifier (measured quantity – photocurrent, Fig. 4.1) and with an external operational amplifier (measured quantity – voltage, Fig. 4.2).

Semiconductor light detectors are also called "square law detectors." This is because the probability of charge generation is proportional to the square value of the field strength of the radiation field illuminating the detector area. Table 4.1 summarizes data on semiconductor light detectors.

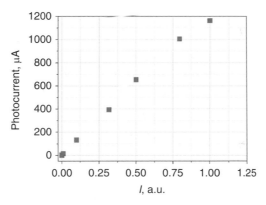

Figure 4.1
Photodiode SD-444-12 in photoconductive mode (light source – green 35 mW cw laser, reverse bias 5 V). The irradiance is controlled by means of neutral filters and is expressed in arbitrary units (a.u.).

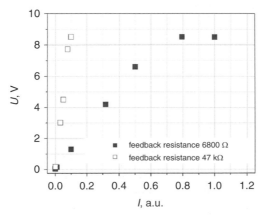

Figure 4.2
Photodiode SD-444-12 in photoconductive mode with operational amplifier (light source – green 35 mW cw laser): detected signal vs. emitted power. The irradiance is controlled by means of neutral filters and is expressed in arbitrary units (a.u.).

Table 4.1 A brief summary of light detectors

| | Output (measured) signal | | |
Light detector	Response linearity	Measured quantity	Comments
Photoresistor	non-linear	photocurrent/ voltage	simple electrical circuit
Photodiode (photovoltaic mode)	non-linear	voltage	simple electrical circuit, does not require external voltage supply
Photodiode (photoconductive mode)	linear	photocurrent/ voltage	simple electrical circuit
Photodiode (photoconductive mode with operational amplifier)	linear	voltage	tunable dynamic diapason

Materials Needed

- Optical table/Optical rail
- Optical attenuators labeled #1, #2, #3, #4
- Light sources (LED, tungsten filament lamp, red cw laser)
- Photodiode OPT101
- Proto-board
- One operational amplifier LM741
- Resistance/capacitance box
- Resistors (10–50 Ω) and two electrolytic capacitors (1.0 µF)
- Two-channel DC voltage supply
- Electrical cables and wires
- Multimeter
- Ruler
- Mechanical holders
- Screws and screwdrivers
- Catalogs of optical companies (Newport, Edmund Optics, Thorlabs etc.)
- Specification sheets for semiconductor light detectors (photodiodes) used

Procedures

Measuring Light Using Photodiodes

(a) Photoconductive (Reverse-Biased) Mode

1. Design the electrical circuit shown in Fig. 4.3.

2. Assemble the optical set-up shown schematically in Fig. 4.4a (the photodiode should be connected in series to the power

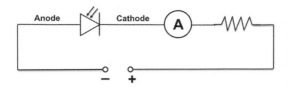

Figure 4.3 Electrical circuit of a light detector based on a photodiode: Photoconductive (reverse-biased) mode of operation.

(a)

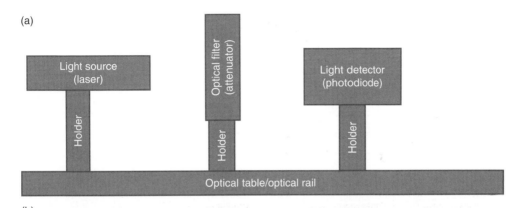

(b)

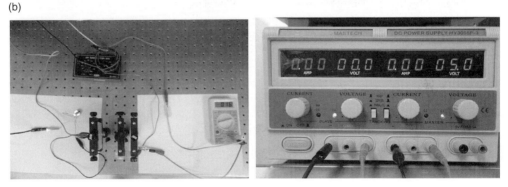

Figure 4.4 (a) Optical set-up used to characterize a light detector based on a photodiode. Use a laser or an LED as the light source (in the case of an LED, refer to Chapter 3 on how to connect an LED to the power supply). (b) Photo of the set-up shown in Fig. 4.4a (left), a two-channel DC power supply feeding an LED and a photodiode (right).

supply and ammeter according to Fig. 4.3). Use a 100 Ω resistor and apply a voltage of 5 V.

3. Open the light input of the photodiode and measure the photocurrent through the photodiode that is generated by daylight.

4. Switch on the light source,[1] and, by placing different optical attenuators (#1, 2, 3, 4), measure the photocurrent through the photodiode versus irradiated light power. Complete the following table.

Table 4.2

Applied DC (V)	Resistance in series (Ω)	Attenuator	Transmittance (%)	Laser power (mW) (if you use a laser as a light source)	Power (a.u.) (if you use an LED as a light source)	Photocurrent (daylight) (μA)	Photocurrent (laser light + daylight) (μA)	Photocurrent (laser light)[*] (μA)
		NO	100		1			
		#1						
5	100	#2						
		#3						
		#4						

[*] Photocurrent (laser light) = Photocurrent (laser light + daylight) − Photocurrent (daylight)

5. Plot the dependence of photocurrent on light power.

(b) Photoconductive (Reverse-Biased) Mode with Operational Amplifier

1. Design an electrical circuit as shown in Fig. 4.5a. Refer to Useful Links at the end of the chapter for the technical characteristics of the operational amplifier LM741.

2. Assemble the optical set-up shown schematically in Fig. 4.4 and pictured in Fig. 4.5c.

3. Close the light input of the photodiode and connect the multimeter probes to the output (V_{out}) of the designed electrical circuit (Fig. 4.5a). The measured value of output voltage should be ~0 V.

4. Open the light input of the photodiode while simultaneously observing changes in the output voltage on the screen of the multimeter.

---

[1] Going forward, we will refer to a laser as a light source when used in optical set-ups, lab text, instructions, and tables. However, an LED may be used equally well, and in many cases may provide better results. By trying both light sources, you may determine your preference for particular tasks.

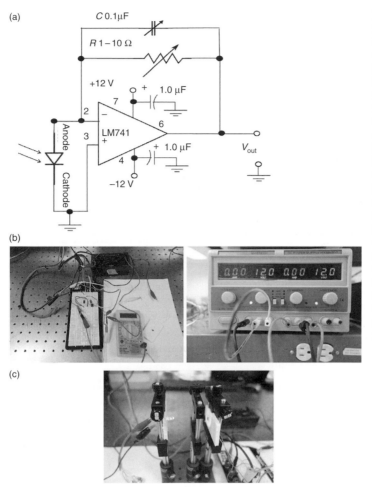

(a)

(b)

(c)

Figure 4.5
(a) Electrical circuit of a light detector based on a photodiode: Photoconductive (reverse-biased) mode of operation with operational amplifier. (b) Photos of schematics shown in Fig. 4.5a (left) and a two channel DC power supply (right). The most challenging moment for students at this stage is how to produce +12V and −12V from the power supply. Work with more experienced colleagues or refer to external literature to ensure understanding of both. (c) Optical set-up.

If you do not see any changes in the output voltage, check that your design matches the electrical circuit shown in Fig. 4.5a. Find and correct any mistakes you may have made.

For Advanced Students: To simplify the diagnostics of the designed electrical circuit, replace the photodiode with a function generator. In this case, you will know everything about the parameters (signal shape, amplitude, and frequency) of the input electrical signal. The output signal shape and frequency should remain the same, but its amplitude should be amplified. The value of the amplification can be tuned by changing the resistance of the feedback resistor (R = 1–10 kΩ in Fig. 4.5a). Remember, the maximum value of the output voltage cannot exceed the voltage of the power supply (in our case 12 V).

5. Switch on the laser. Direct the laser beam into the light input of the photodiode and measure the output voltage (V_{out}). Change the resistance of the feedback resistor to identify a value at which the output voltage is closer to ~10 V. This value of the resistor provides the recommended amplification and ensures that the dynamical diapason of the amplifier is used to its full extent. Record your findings in the table below.

Table 4.3

Input light signal	Emitted light power is ~25 mW (for red lasers)								
Feedback resistance (ohm)	100	220	470	1000	1500	2200	4700	6800	10000
Output voltage (V)									

6. Plot output voltage vs. feedback resistance and find the optimal value of feedback resistance for measuring the light power of the light source (red laser). For the optimal feedback resistance found, measure the output voltage caused by daylight.

7. Switch on the laser and, by placing different optical attenuators (#1, 2, 3, 4), measure the dependence of the output voltage on irradiated light power. Complete the following table.

Table 4.4

Attenuator	Transmittance (%)	Power (mW)	Power (a.u.)	Output voltage (daylight) (V)	Output voltage (laser light + daylight) (mV)	Output voltage (laser light)* (V)
NO	100	25	1			
#1						
#2						
#3						
#4						

* Output voltage (laser light) = Output voltage (laser light + daylight) − Output voltage (daylight)

8. Plot the dependence of output voltage on light power.

Evaluation and Review Questions

1. Describe the advantages of a reverse-biased photodiode with an external amplifier over a reverse-biased photodiode (without an amplifier).
2. Assume that you connect a forward-biased photodiode to the electrical circuit shown in Figs. 4.3 and 4.5. Will the photodiode detect light in each case?
3. Describe the procedure used to calibrate a linear photodetector based on a photodiode. Assume that you have in your possession a light source of known output power measured with high accuracy.

For Further Investigation

Inverse Square Law

1. Design the electrical circuit shown in Fig. 4.6.

2. Assemble the optical set-up shown schematically in Fig. 4.7. For light detection, use any of the devices and circuitry studied up to this point, such as a photoresistor or photodiode. The distance between the emitting surface of the incandescent lamp and the light input of the photodiode should be ~2 cm.

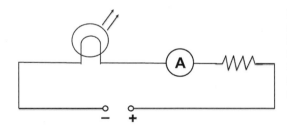

Figure 4.6 Electrical circuit: Light source (incandescent lamp), ammeter, resistor (10 Ω), and voltage supply (5–10 V) connected in series.

Figure 4.7 Optical set-up.

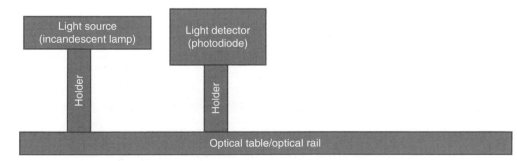

3. Switch on the power supply, direct the light from the incandescent lamp into the light input of the photodiode, and measure the output voltage (V_{out}). Adjust the output voltage close to ~10 V by changing the resistance of the feedback resistor.

4. By changing the distance between the light source and the light detectors, measure the dependence of irradiance on distance. Complete the following table.

Table 4.5

Distance (cm)	(Distance)$^{-2}$ (cm^{-2})	Output signal (V)
2		
4		
6		
8		
10		
20		

5. Plot the dependence of output signal on (distance)$^{-2}$. Explain the character of the dependence.

Further Reading

General

F. L. Pedrotti, S. J. L. Pedrotti, L. M. Pedrotti, *Introduction to Optics*, 3rd edition, Upper Saddle River, NJ: Pearson Prentice Hall, 2007

Specialized

Handbook of Optics, W. G. Driscoll (editor), W. Vaughan (associate editor), New York: McGraw-Hill, 1978

K. F. Brennan, *The Physics of Semiconductors with Applications to Optoelectronic Devices*, New York: Cambridge University Press, 2003

L. Desmarais, *Applied Electro-Optics*, Upper Saddle River, NJ: Prentice Hall, 1997

G. Rieke, *Detection of Light from the Ultraviolet to the Submillimeter,* 2nd edition, New York: Cambridge University Press, 2003

B. G. Yacobi, *Semiconductor Materials: An Introduction to Basic Principles*, New York: Kluwer Academic/Plenum Publishers, 2003

Chapters 2 and 3 of this book

Catalogs and Web Resources

www.coherent.com/ - Coherent
www.cvimellesgriot.com -— CVI Melles Griot
www.deltron.com/ - Del-Tron Precsiscion Inc.
www.edmundoptics.com/ - Edmund Optics
www.newport.com/ - Newport Corporation
www.thorlabs.com/ - Thorlabs
www.ti.com/ - Texas Instruments

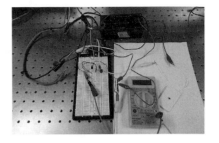

Useful Links

LM741 Operational Amplifier
Datasheet: www.ti.com/lit/ds/symlink/lm741.pdf

5 Basic Laws of Geometrical Optics: Experimental Verification

Objectives

1. State and experimentally verify the law of transmission.
2. State and experimentally verify the law of reflection.
3. Measure reflectance of a mirror.
4. State and experimentally verify Snell's law of refraction.
5. Measure the index of refraction of a solid (polymeric sheet or glass) material using Snell's law of refraction.
6. Examine beam displacers based on plane parallel plates.
7. Experimentally observe the process of total internal reflection.
8. Determine the refractive index of a prism material by measuring the angle of total internal reflection.
9. Formulate the limitations of geometrical optics.

Background

Many optical devices (glasses, car mirrors, telescopes, projectors, etc.) can be fabricated and understood using the approach of geometrical optics. Within the approximation represented by geometrical optics, light travels in straight lines, or rays. The idea of light rays traveling in straight lines through space is accurate as long as the wavelength of the radiation is much smaller than the windows, passages, and holes that can restrict the path of the light. When this is not true, the phenomenon of diffraction must be considered, and this is the domain of physical optics.

Geometrical optics is based on three basic laws:

1. The law of rectilinear propagation (transmission). In a region of constant refractive index n, light travels in a straight line.

2. The law of reflection. When a ray of light is reflected at an interface dividing two optical media, the reflected ray remains

Basic Laws of Geometrical Optics: Experimental Verification

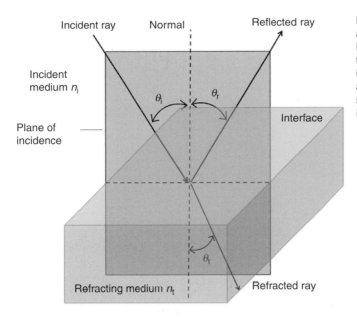

Figure 5.1 Reflection and refraction at an interface between two optical media: Incident, reflected, and refracted rays are shown in the plane of incidence.

within the plane of incidence, and the angle of reflection θ_r equals the angle of incidence θ_i. The plane of incidence is the plane containing the incident ray and the surface normal at the point of incidence (Fig. 5.1).

Mathematically, the law of reflection is very simple: $\theta_r = \theta_i$

3. The law of refraction (Snell's law). When a ray of light is refracted at an interface dividing two transparent media, the transmitted ray remains within the plane of incidence and the sine of the angle of refraction θ_t is directly proportional to the sine of the angle of incidence θ_i (Fig. 5.1).

Mathematically, the law of refraction can be written as follows: $n_i \sin \theta_i = n_t \sin \theta_t$, where n_i and n_t are the refractive indices of the incident and refracting media, respectively. The refractive index n is the ratio of the speed of light in a vacuum c divided by the speed of light in the medium v: $n = c/v$.

A special case must be considered if the refractive index of the incident medium is greater than that of the transmitting medium, $n_i > n_t$. In this case, the law of refraction can be rewritten as $\sin \theta_t = (n_i/n_t) \sin \theta_i$ and we then find a *critical angle* $\theta_i = \theta_c$, where $\sin \theta_c = n_t/n_i$. For the critical angle, $\sin \theta_t = 1$, the transmitted ray travels perpendicularly to the normal (parallel to the interface), as shown in Fig. 5.2. For incident angles greater than the critical angle, no light is transmitted. Instead, the light is totally reflected into the incident medium (Fig. 5.2). This effect is called *total*

Figure 5.2 Refraction of light rays at angles near or at the critical angle.

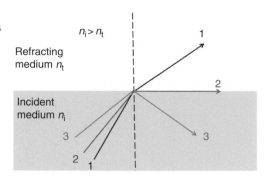

internal reflection. It plays an important role in the life of modern digital society, specifically in optical fibers, the heart of optical communications.

In Fig. 5.2, ray 1 is transmitted into a refracting medium; ray 2 is incident at the critical angle θ_c, and as a result the transmitted light travels along the surface of the interface; ray 3 is incident at an angle greater than θ_c and is totally internally reflected.

Materials Needed

- Optical table or optical rail
- Light sources (red cw laser)
- Photodetector based on Photodiode OPT101 and operational amplifier
- Ruler
- Mechanical holders
- Screws and screwdrivers
- Circular protractor
- Squared graph paper
- Screen
- Needles
- Holding surface for mirror and prism
- Mirror
- Prisms (right-angle prism, equal-angle prism)

Procedures

Experimental Verification of the Law of Linear Propagation (Transmission)

Although our everyday experience (for example, rays of sunlight pouring in through the window of your room) convinces us that light travels in straight lines, it is interesting to verify the law of rectilinear propagation (transmission) in the lab. For this purpose we will use a laser beam – perhaps the best actual approximation to a ray of light (for the sake of simplicity, as long as we are concerned with geometrical optics, the spatial (mode) structure of the laser beam will not be discussed here).

1. Draw a straight line on a sheet of a paper.

Basic Laws of Geometrical Optics: Experimental Verification

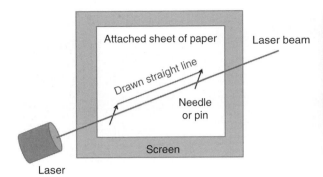

Figure 5.3 Optical set-up to verify the law of rectilinear propagation transmission.

2. Attach the sheet of a paper to a screen using adhesive tape or pins.

3. Attach a needle to each end of the straight line.

4. Switch on the laser and direct the laser beam parallel to the drawn straight line as shown in Fig. 5.3.

5. Draw conclusions about the validity of the law of rectilinear propagation transmission.

Experimental Verification of the Law of Reflection

1. Identify a mirror available in your optics lab. What type of mirror do you have?

2. Assemble the optical set-up shown in Fig. 5.4a and measure the output power of the laser using a photodetector based on a photodiode and operational amplifier.

3. Assemble the optical set-up shown schematically in Fig. 5.4b.

4. Measure the angle of incidence and the angle of reflection using a circular protractor. Additionally, measure the power of the reflected laser beam. In your case, the detector is not graduated in power units. As a result, the power of the beam will be measured using arbitrary units.

     To measure angles using a circular protractor, trace the laser beam with needles. This technique is based on the well-known geometrical axiom that only two dots determine a straight line. Since the laser beam is a straight line (in terms of geometrical optics), attach two needles to the holding surface and the laser beam should pass through these two needles. After doing this, draw a straight line which will be the direction of the incident light beam (Fig. 5.5). In an analogous way, trace the reflected laser beam and measure the angles of incidence and reflection.

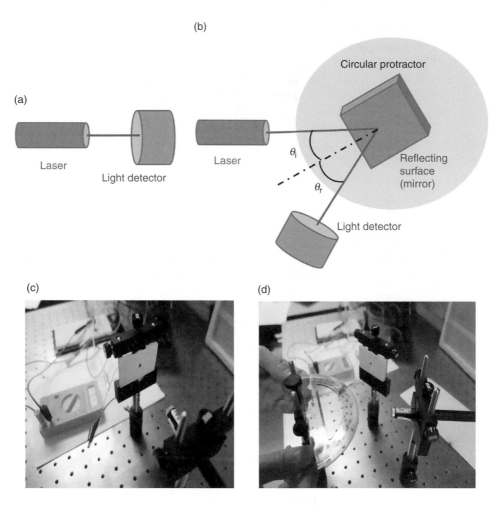

(b)

Circular protractor

(a)

Laser

Light detector

Laser

θ_i

Reflecting surface (mirror)

θ_r

Light detector

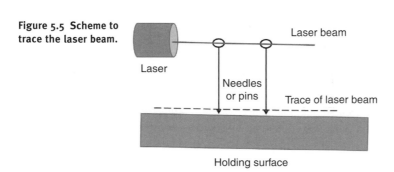

(c)   (d)

Figure 5.4 Optical set-up to verify the law of reflection.

Figure 5.5 Scheme to trace the laser beam.

Laser beam

Laser

Needles or pins

Trace of laser beam

Holding surface

Basic Laws of Geometrical Optics: Experimental Verification

5. Complete Table 5.1.

   Table 5.1

Angle of incidence, θ_i	Angle of reflection, θ_r	Incident light power (a.u.)	Reflected light power (a.u.)	Reflectance of surface[*] (%)
$0°$				
$15°$				
$30°$				
$45°$				
$60°$				
$75°$				
$90°$				

   [*] Reflectance of a surface is defined as the ratio of the reflected light power to the incident light power.

6. Draw conclusions about the validity of the law of reflection.

7. Plot the reflectance of the surface vs. angle of incidence. Does geometrical optics explain the plotted curve?

Experimental Verification of the Law of Refraction (Snell's Law)

1. Identify a plane parallel plate available in your optics lab.

2. Assemble the optical set-up shown in Fig. 5.6a. Refer also to the pictures shown in Fig. 5.6b.

3. Measure the angle of incidence, angle of reflection, and the angle of refraction using a circular protractor. Use needles to trace the incident, reflected, and refracted laser beams. Mark the position of the plane parallel plate on a sheet of paper and mark the dots along which the laser beam enters and exits the plane parallel plate.

4. Change the angle of incidence and complete Table 5.2.

   Table 5.2

Angle of incidence, θ_i	Angle of reflection, θ_r	Angle of refraction, θ_t	Refractive index, n[*]
$0°$			N/A
$15°$			

Table 5.2 (cont.)			
Angle of incidence, θ_i	Angle of reflection, θ_r	Angle of refraction, θ_t	Refractive index, n^*
30°			
45°			
60°			
75°			
90°			

$^*n = \sin\theta_i / \sin\theta_t$

(a)

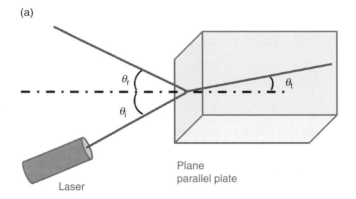

(b)

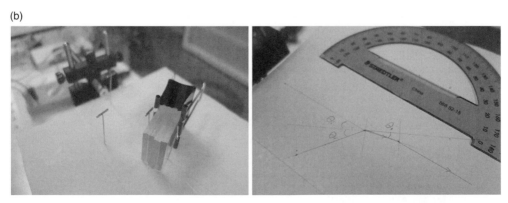

Figure 5.6 (a) Optical set-up to verify the law of refraction. (b) On a sheet of paper, mark the position of the plane parallel plate and the dots along which the laser beam enters and exits the plane parallel plate.

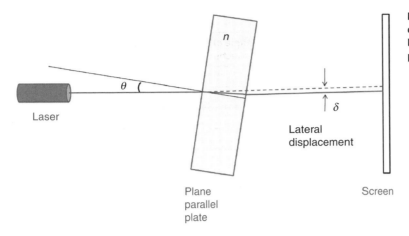

Figure 5.7 Lateral displacement of a laser beam by a plane parallel plate.

5. Analyze the data obtained and conclude on the validity of the law of refraction.

Application of Plane Parallel Plate Laser Beam Displacer

1. Assemble the optical set-up shown in Fig. 5.7.

2. Change the angle θ from 0° to 60°, and for each angle measure the displacement of the laser beam. Complete Table 5.3.

Table 5.3

Angle of incidence, θ_i	Measured displacement, δ (mm)	Refractive index, n
0°		N/A
15°		
30°		
45°		
60°		

To calculate the refractive index, use the following expression:
$\delta = d \sin \theta \left[1 - \cos \theta / \sqrt{n^2 - \sin^2 \theta} \right]$, where d is the thickness of the plane parallel plate in the direction of light propagation, n is the refractive index of the plate, and θ is the angle of incidence.

$$n^2 = (\sin \theta)^2 + \frac{(\cos \theta)^2}{\left(1 - \frac{\delta}{d \sin \theta} \right)^2}$$

Figure 5.8 Total internal reflection in prisms.

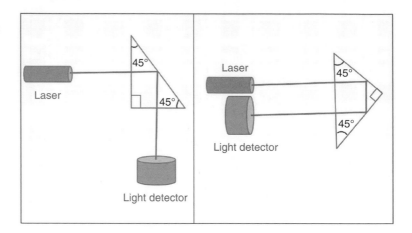

3. Analyze the results obtained. For different angles of incidence, explain why you found slightly different values of refractive index.

4. Estimate the experimental error of the measured value of the refractive index.

Total Internal Reflection from a Prism: Prism as a Mirror

1. Identify a right-angle prism in your optics lab.

2. Assemble the optical set-up shown in Fig. 5.8.

3. Prove that you observe total internal reflection.

4. Measure the reflectance of the prism in both the cases shown in Fig. 5.8. Compare the reflectance of a prism and a regular mirror and complete Table 5.4.

Table 5.4

	Prism	Mirror
Reflectance		

5. Describe the advantages and disadvantages of a prism used as a mirror reflector instead of a regular mirror.

Refractive Index of a Prism Material Found by Measuring the Angle of Total Internal Reflection

1. Select the optical set-up shown in Fig. 5.9a.

2. Switch on a laser and direct the laser beam onto the surface of a prism as shown in Fig. 5.9a. Rotate the prism carefully until the

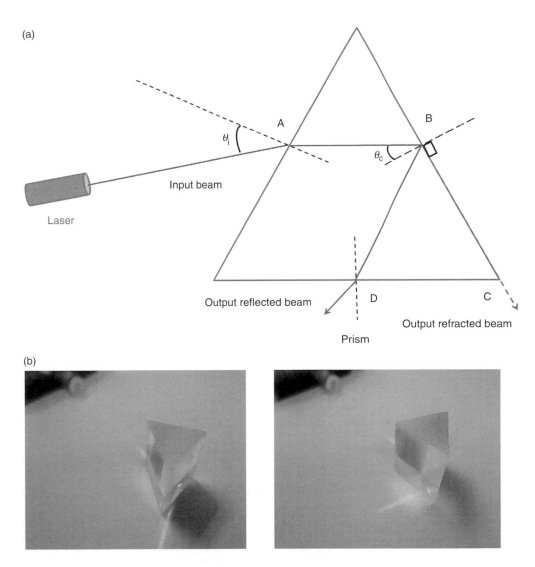

(a)

Laser

Input beam

θ_i

A

B

θ_c

Output reflected beam

D

C

Output refracted beam

Prism

(b)

Figure 5.9 **(a) Scheme to determine refractive index of a prism using total internal reflection (for clarity, laser beams reflected at interfaces are not shown). (b) There is no total internal reflection in the picture on the left, and there is total internal reflection in the picture on the right.**

output refracted beam comes out along the edge of the surface (BC in Fig 5.9a), creating conditions of total internal reflection. Refer to Fig. 5.9b for images.

3. Trace the laser beams shown in Fig. 5.9a using needles and, by drawing straight lines corresponding to each laser beam, draw angle ABD.

4. Measure angle ABD using a protractor.

5. Calculate the refractive index of the prism material using the following expression for total internal reflection:

$$\sin \theta_c = \sin \left(\frac{\text{ABD}}{2} \right) = \frac{1}{n_{\text{prism}}}$$

$$n_{\text{prism}} =$$

6. Compare the values obtained for the refractive index of the prism and the plane parallel plate.

Evaluation and Review Questions

1. Derive an expression for the lateral displacement of a laser beam passing through a plane parallel plate.
2. Is it possible to measure the dispersion of light (wavelength dependence of refractive index) using the method based on displacement of a laser beam passing through a plane parallel plate? Hint: Estimate the accuracy of this method and compare it with the typical deviation of refractive index due to wavelength changes for regular glass (find these data in the literature or online).
3. Derive an expression for the minimum angle of deviation of a prism.

For Further Investigation

You can fabricate your own mirrors, plane parallel plates, and prisms using inexpensive rough materials (sheet of aluminum foil, glass, and polymers such as Plexiglas).

For example, aluminum foil can be used to make a simple mirror; plane parallel plates and prisms can be fabricated using polymer or regular glass.

Further Reading

General

E. Hecht, *Optics*, 4th edition, San Francisco, CA: Addison-Wesley, 2001

F. L. Pedrotti, S. J. L. Pedrotti, L. M. Pedrotti, *Introduction to Optics*, 3rd edition, Upper Saddle River, NJ: Pearson Prentice Hall, 2007

Fundamentals of Photonics, C. Roychoudhuri (editor), Bellingham, WA: SPIE Press, 2008. This tutorial text is available via a free download using link http://spie.org/x17229.xml.

Specialized

Handbook of Optics, W. G. Driscoll (editor), W. Vaughan (associate editor), New York: McGraw-Hill, 1978

D. C. O'Shea, *Elements of Modern Optical Design*, New York: Wiley, 1985

Chapter 2 of this book

6 Converging and Diverging Thin Lenses

Objectives

1. Develop basic skills to identify, classify, and handle thin lenses.
2. Develop basic skills to characterize a lens "at a glance."
3. Measure focal lengths of converging lenses using different methods: Conjugate-foci method, unit magnification method, and Bessel method.
4. Determine focal lengths of thin negative (diverging) lenses using parallel light beams and using an auxiliary converging lens.
5. Formulate the limitations of the studied experimental techniques.

Background

A lens is used as an optical component in almost all optical devices. A lens is a piece of optical material (transparent refracting medium, for example glass or polymer; for details see Further Reading) with spherically shaped surfaces on the front and back. A ray incident on the lens refracts at the front surface, propagates through the lens, and refracts again at the rear surface (according to the basic laws of geometrical optics). Figure 6.1 shows a lens refracting rays from an object AB to form an image A′B′, where R_1 and R_2 are the radii of curvature of the lens surfaces, d is the lens axial thickness, n is the refractive index of the lens material, s is the object distance, and s' is the image distance. The lens shown in Fig. 6.1 is symmetrical about a line, called the *optical axis*.

If the axial thickness d of a lens is small compared with the radii of curvature of its surfaces ($d \ll R_{1,2}$), such a lens can be treated as a *thin* lens. Thin lenses can be *converging* or *diverging* as shown in Fig. 6.2a,b, respectively. Converging, or positive, lenses are thicker in the middle than

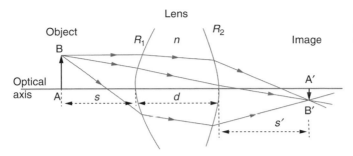

Figure 6.1 Light ray propagation through a lens.

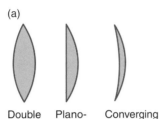

(a)

Double convex   Plano-convex   Converging meniscus

(b)

Double concave   Plano-concave   Diverging meniscus

Figure 6.2 (a) Converging lenses. (b) Diverging lenses.

at the edges. Diverging, or negative, lenses are thinner in the middle than at the edges. Converging lenses cause parallel rays passing through them to bend toward one another. Diverging lenses cause parallel rays passing through them to spread as they leave the lens.

Thin lenses and their combinations are widely used to form images of objects. For the so-called paraxial approximation (first-order or Gaussian optics, assuming $\sin \alpha \sim \alpha$), when light rays trace close to the optical axis of the lens, the following *thin-lens equation* (6.1) can be derived:

$$\frac{1}{s} + \frac{1}{s'} = \frac{1}{f} \qquad (6.1)$$

where s is the object distance, s' is the image distance, and f is the focal length of the lens. The focal length of the thin lens is defined as the image distance for an object at infinity or the object distance for the image at infinity. The expression for the focal length of the lens (or *lensmaker's equation*) (6.2) can also be found by using the paraxial approximation (it is supposed that the lens is used in air; for details see the literature):

$$\frac{1}{f} = (n-1)\left(\frac{1}{R_1} - \frac{1}{R_2}\right) \qquad (6.2)$$

63

The lensmaker's equation (6.2) predicts the focal length of a lens fabricated with a given refractive index n and radii of curvature R_1, R_2 used in air. In the case when a lens is used in a medium of refractive index n_0, the lensmaker's equation can be written as follows:

$$\frac{1}{f} = \left(\frac{n - n_0}{n_0}\right)\left(\frac{1}{R_1} - \frac{1}{R_2}\right)$$ (6.3)

The value $1/f$ is the refracting power P of the lens: $P = 1/f$. The unit of refracting power is *reciprocal length*. When the lengths are measured in meters, their reciprocals are said to have units of *diopters* (D).

It should be noted that both the thin-lens equation and the lensmaker's equation use the so-called *sign conventions*:

- Lenses are assumed to be axially symmetrical.

- Light rays are assumed to progress from left to right.

- Lenses have positive power if they converge light (i.e. power is positive for a converging lens and negative for a diverging lens).

- Distances upward (or to the right) are positive.

- Object distance to the left of the lens is positive.

- Object distance to the right of the lens is negative.

- Image distance to the right of the lens is positive.

- Image distance to the left of the lens is negative.

- The refractive index is considered positive when the light ray travels in the normal left-to-right direction.

- Angles of incidence, refraction, and reflection are positive if the ray is rotated clockwise to reach the normal to the surface.

- Slope angles are positive if the ray is rotated counterclockwise to reach the axes (see also *Modern Optical Engineering* by W. J. Smith).

- Radii and curvatures are positive when the center of curvature is to the right of the surface.

A simple geometrical approach is used to sketch the image of an object formed by a thin lens. Figure 6.3 shows diagrams of image formation by converging and diverging lenses.

The performance of any thin lens is determined by (1) focal length, (2) diameter, (3) absorption of the glass, and (4) aberrations. All of these parameters can be determined experimentally.

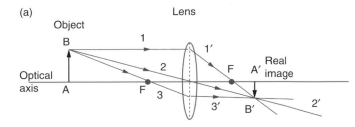

(a)

Figure 6.3 Ray diagrams for image formation by (a) converging and (b) diverging lenses.

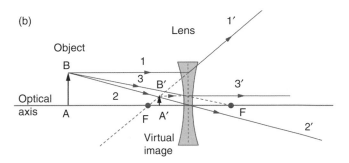

(b)

Procedures

Wear gloves or finger cots to handle optical elements (lenses).

Perform a Brief (At-a-Glance) Lens Examination

As a rule, a thin lens is a simple component of a more complicated optical experiment. There are a few useful, practical rules about how to identify the right lens at a glance, without sophisticated and time-consuming measurements. Below, we describe the procedure.

1. Measure the thicknesses of all lenses available in your optics kit using calipers or a ruler. Identify thin lenses for future experiments. (Hint: From a practical point of view, a lens can be considered as a "thin" lens when its thickness is smaller than a few (~3–5) millimeters). Complete the following table.

Materials Needed

- Optical table or optical rail
- Light source (red cw laser)
- Object (arrow)
- Ruler
- Mechanical holders
- Screws and screwdrivers
- Calipers and/or spherometer
- Squared graph paper
- Screen
- Needles
- Holding surface for mirror and prism
- Mirror
- Thin lenses: Plano-concave, plano-convex

Table 6.1

Available lens	Lens thickness (mm)	Type of lens	
		Thick lens	Thin lens
Lens 1			
Lens 2			
Lens 3			
Lens 4			
Lens 5			

It should be pointed out that the same lens can be "thin" or "thick" depending on the particular conditions of the experiment. Think about this.

2. Analyze the lens shape of the thin lenses selected in the previous section. Make the lens classification and complete the following table.

Table 6.2

	Common types of thin lenses					
	Converging lenses			Diverging lenses		
Lens to analyze	Double convex	Plano-convex	Converging meniscus	Double concave	Plano-concave	Diverging meniscus
#1						
#2						
#3						
#4						

The type of a "thin" lens – converging or diverging – can be determined by visual analysis of the lens shape. The situation becomes confusing when visual analysis cannot help you identify the lens type. In such a case, try to use the lens as a magnifier: read texts through the lens. The lens should be very close to the text (the distance between the text and the lens should be smaller than the focal length. Since you do not know the focal length, start with the smallest distance). In the case of a converging lens, text will be magnified, and in the case of a diverging lens, it will be minified.

Converging and Diverging Thin Lenses

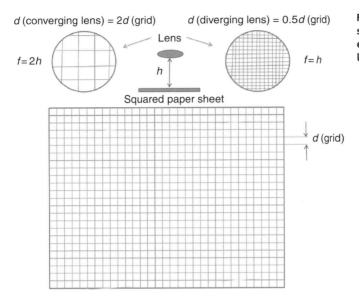

d (converging lens) = 2*d* (grid)   *d* (diverging lens) = 0.5*d* (grid)

f = 2*h*

Lens

h

Squared paper sheet

f = *h*

d (grid)

Figure 6.4 Set-up of a simple technique to estimate the focal length of lenses.

3. Place the lens on the squared paper sheet (Fig. 6.4). You will notice visible changes in the size of the grid squares by comparing areas of the paper covered and not covered by the lens.

4. Take the lens and move it up in order to increase the distance *h* between the paper surface and the lens. Simultaneously, observe changes in the size of the grid squares observed with the lens. Continue to increase the distance between the lens and the squared paper until the observed period of the grid is double-changed (increased by a factor of 2 for the converging lens and decreased by a factor of 2 for the divergence lens, as shown in Fig. 6.4). Measure the distance *h* between the paper surface and the lens for each lens. As can be shown via direct geometrical ray tracing, such a distance *h* relates to the focal length *f* of lenses by simple expressions (derive them):

$$f = 2h \text{ for converging lens,}$$

$$f = h \text{ for diverging lens.}$$

5. Measure the diameters of the thin lenses using calipers. For each lens, calculate the *f-number* (the ratio of the focal length to the diameter of the lens) and the refracting power *P* given by $1/f$ (*f* is focal length). Complete the following table.

Table 6.3

Lens	Focal length* (mm)	Refracting power (D = m^{-1})	Lens diameter (mm)	f-number**
#1				
#2				
#3				
#4				

*Remember the sign conventions: focal length is positive for a converging lens and negative for a diverging lens.
**A convenient way to express f-number is as a ratio instead of a decimal number; for example 1:2 instead of 0.5.

6. Develop your own express method to characterize unknown lenses based on the procedures described above. Estimate the time you will need to complete an "at -a-glance" characterization of one lens.

Determination of Focal Length of Thin Positive (Converging) Lenses

Conjugate-Foci Method

1. Assemble the optical set-up shown in Fig. 6.5.

2. Switch on the light source. Put the object (arrow) at eight different positions with respect to the lens. For each position, determine and measure the corresponding distances s and s'. To find s', change the position of the screen until you see a well-defined image of the object on the screen (Fig. 6.5b).

3. Use the thin-lens equation, $\frac{1}{f} = \frac{1}{s} + \frac{1}{s'}$, to calculate the focal length f.

4. Repeat the measurements for other thin converging lenses and complete the following table.

Table 6.4

Lens 1			Lens 2		
s (mm)	s' (mm)	f (mm)	s (mm)	s' (mm)	f (mm)

Converging and Diverging Thin Lenses

(a)

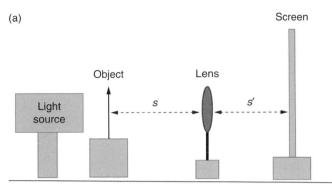

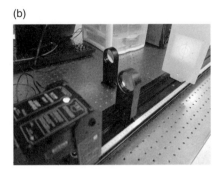

Optical table or rail

Figure 6.5 Optical set-up to measure the focal length of a thin converging lens: Conjugate-foci method.

(b)

5. Calculate the average values of the focal length for two lenses.

$\langle f \rangle =$ $\qquad$ $\langle f \rangle =$

Unit Magnification Method

1. Assemble the optical set-up shown in Fig. 6.6.

2. Switch on the light source. Place the object and screen symmetrically with respect to the lens, and by changing distances s and s', arrange conditions for unit magnification as shown in Fig. 6.6.

3. For the conditions found, measure the distance D between the object and the screen and calculate the focal length of the lens using this expression:

$$f = \frac{D}{4}$$

4. Change the lens and repeat all the required experimental steps to find the focal length.

5. Write down the results obtained for two lenses:

   $f =$

   $f =$

6. Discuss the limitations of this method. (Hint: Is it possible to measure very small or very large focal lengths using this method?)

Bessel Method

1. Assemble the optical set-up shown in Fig. 6.7.

2. Switch on the light source. Select the appropriate constant distance L between the object and the screen, and move the lens along the optical axis between the fixed object and fixed image screen. Find two positions of the lens for which an image is in focus on the screen, magnified in one case, and reduced in the other.

Figure 6.6 Optical set-up to measure focal length of thin converging lens: Unit magnification method.

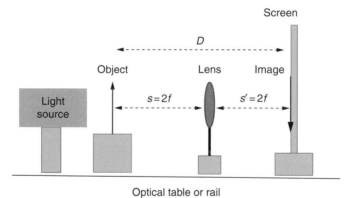

Figure 6.7 Optical set-up to measure focal length of thin converging lens: Bessel method.

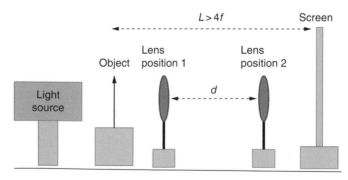

3. Measure the distance d between the two lens positions, as shown in Fig. 6.7.

4. Calculate the focal length of the lens using the following expression: $f = \frac{L^2 - d^2}{4L}$

5. Change the lens and repeat all the required experimental steps to find the focal length.

6. Complete the following table.

Table 6.5

	L (mm)	d (mm)	f (mm)
Lens 1			
Lens 2			

7. Discuss the limitations of this method.

Determination of the Focal Length of Thin Negative (Diverging) Lenses Using Parallel Light Beams

We use a laser beam as a parallel light beam. A real laser beam has finite angular divergence. To obtain good results, the output light beam angular divergence, which is caused by the diverging lens, should be much greater than the intrinsic angular divergence of the laser beam.

1. Assemble the optical set-up shown in Fig. 6.8.

2. Put the screen at eight different positions with respect to the diverging lens (Fig. 6.8 shows two positions). For each such position, measure distance x_i and the width of the transmitted laser beam y_i. Complete the following table.

Table 6.6

x_i (mm)							
y_i (mm)							

3. Plot y_i vs. x_i as shown in Fig. 6.9. The expected dependence is
$$y = 2\,\text{const}(x + f) = C(x + f)$$

4. By fitting the plotted dependence with a straight line, determine the focal length of a diverging lens, as shown in Fig. 6.9.

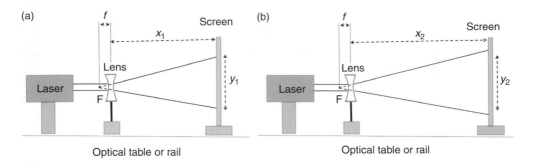

Optical table or rail

Figure 6.8 Optical set-up to measure focal length of a thin diverging lens using a laser beam.

Figure 6.9 Determination of focal length by fitting experimental data with a linear dependence.

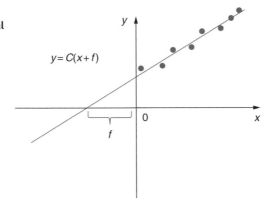

5. Write down the result obtained for a diverging lens:

   $f =$

6. Draw a conclusion about the limitations of the considered experimental method.

Determination of the Focal Length of Thin Negative (Diverging) Lenses Using an Auxiliary Converging Lens

1. Assemble the optical set-up shown in Fig. 6.10.

2. Place a diverging lens in direct contact with a converging lens. *Make sure that the resulting combination of lenses acts as a converging lens.* It's important to use a converging lens of known focal length.

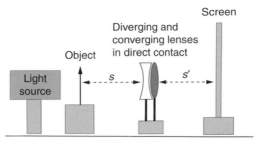

Screen

Figure 6.10 Diverging and converging lenses in direct contact.

3. Since the combination of converging and diverging lenses acts as a single converging lens, apply any method provided thus far to determine the focal length of the converging lens.

4. Measure the focal length of the combination of converging and diverging lenses by using at least two different techniques. Complete the following table.

Table 6.7

Focal length of a single converging lens, f_1 (mm)	Focal length of a combination of diverging and converging lenses, f (mm)	Experimental method used to determine focal length f	Focal length of a diverging lens, f_2 (mm)*

* To determine f, use the following expression for systems of two lenses in direct contact: $\frac{1}{f} = \frac{1}{f_1} + \frac{1}{f_2}$

5. Discuss the limitations of this method.

Evaluation and Review Questions

1. Think about glass materials. How can one distinguish polymeric lenses from glass lenses? (Hint: Compare their basic physical properties, which can easily be tested. For example, look at density, hardness, optical uniformity, melting temperature, etc.)

2. Derive the thin-lens equation.

3. Use a lens as a magnifier and derive an expression for the linear magnification and the angular magnification.

4. Draw images of an arrow for the converging and diverging lenses in the following cases:

   (a) $0 < s < f, s = f/2$
   (b) $s = f$
   (c) $f < s < 2f, s = 1.5f$
   (d) $s = 2f$
   (e) $s > 2f, s = 3f$

5. Using data from procedure 2 above (conjugate-foci method), plot $1/s'$ vs. $1/s$. Determine the focal length by fitting the plotted experimental data with linear dependence. Compare the obtained results with the results based on direct calculations using the thin-lens equation. Estimate errors for both cases.

6. Derive the expression used in the Bessel method: $f = \frac{L^2 - d^2}{4L}$.

7. Develop a procedure to determine the refractive index of lenses. (Hint: Use the lensmaker's equation $1/f = (n_l - 1)[(1/R_1) - (1/R_2)]$. Measure the focal length and radii of curvature. Measure the radii of curvature R_1 and R_2 using calipers or a special tool, the spherometer. After identifying f, R_1, and R_2, calculate the refractive index of the lens using the lensmaker's equation.

8. Develop a procedure for how to apply the "parallel light beam" method to determine the focal length of a converging lens.

9. Can the "combination of lenses" method be applied to determine the focal length of converging lenses?

Conclusion

Summarize the results obtained and complete the following table.

Table 6.8

Method used to determine focal length	Converging lenses		Diverging lenses	
	Lens 1	Lens 2	Lens 3	Lens 4
Conjugate-foci method			N/A	N/A
Unit magnification method			N/A	N/A
Bessel method			N/A	N/A
Parallel light beam method				
Combination of lenses method				

For Further Investigation

1. **Experimentally verify *Abbe's method* to measure the focal length of a thin converging lens**

   Let us assume the image of an object is formed on a screen by a converging lens. Fixing the lens during the experiment, move both the object and the image screen to new positions until the image screen again receives a focused image. If S_1 and S_2 are the object positions for two cases, and if the corresponding transverse magnifications of the image are M_1 and M_2, respectively, it can be shown that the focal length of the lens is given by the following expression: $f = \dfrac{(S_2 - S_1)}{\left(\dfrac{1}{M_1} - \dfrac{1}{M_2}\right)}$.

   Derive the expression shown above and design an optical set-up to experimentally verify this method.

2. **Reflection auto-collimation nodal-point method to measure the focal length of a converging lens**

   Design the optical set-up shown in Fig. 6.11 and develop procedures to determine the focal length of a converging lens.

3. Make a positive converging plano-convex lens using the "dropping-wetting" technique. Place a drop of liquid glue on a flat transparent polymeric or glass surface and wait until the drop becomes solid. Using a glass cutter or scissors (depending on the material used for the flat surface), make a converging plano-convex lens.

4. Make a cylindrical lens using polymeric sheet as the lens surface material, glue as the lens filling component, and metallic wire as a frame. Examine the basic properties of such lenses: Diameter, radius of curvature, focal length.

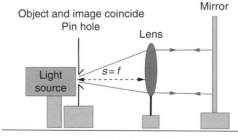

Figure 6.11 Optical set-up to measure focal length of a thin converging lens: Reflection auto-collimation nodal-point method.

Further Reading

General

E. Hecht, *Optics*, 4th edition, San Francisco, CA: Addison-Wesley, 2001

F. L. Pedrotti, S. J. L. Pedrotti, L. M. Pedrotti, *Introduction to Optics*, 3rd edition, Upper Saddle River, NJ: Pearson Prentice Hall, 2007

Fundamentals of Photonics, C. Roychoudhuri (editor), Bellingham, WA: SPIE Press, 2008. This tutorial text is available via a free download using link http://spie.org/x17229.xml.

Specialized

Handbook of Optics, W. G. Driscoll (editor), W. Vaughan (associate editor), New York: McGraw-Hill, 1978

D. C. O'Shea, *Elements of Modern Optical Design*, New York: Wiley, 1985

W. J. Smith, *Modern Optical Engineering*, 3rd edition, New York: McGraw Hill, 2000

A. F. Wagner, *Experimental Optics*, New York: John Wiley & Sons, 1929

Chapters 2 and 5 of this book

Catalogs and Web Resources

spie.org/x17229.xml - SPIE
www.cvimellesgriot.com - CVI Melles Griot
www.edmundoptics.com/ - Edmund Optics
www.newport.com/ - Newport Corporation
www.thorlabs.com/ - Thorlabs

Thick Lenses

7

Background

If the axial thickness d of a lens is not neglected, we treat the lens as being a *thick* lens. This means that the lens thickness is comparable to the radii of curvature of its surfaces ($d \approx R_{1,2}$). In the paraxial limit (Gaussian or first-order optics utilizing approximation $\sin \alpha \approx \alpha$), a thick lens can be described in terms of *cardinal points*. This description is general and can be applied to more complex optical systems consisting of different lenses.

Let us consider the thick lens shown schematically in Fig. 7.1. There are six cardinal points on the optical axis of a thick lens, from which its imaging properties can be deduced. Slightly curved surfaces (which can be considered as planes in the paraxial approximation) normal to the optical axis at cardinal points are called the *cardinal planes*.

The cardinal points consist of the first (front) and second (back) *focal points* (F_1 and F_2); the first (front) and second (back) *principal points* (H_1 and H_2); and the first (front) and second (back) *nodal points* (N_1 and N_2).

The first and second focal points can be defined as the images created from objects located at $-\infty$ for F_1 and $+\infty$ for F_2. In other words,

Figure 7.1 Cardinal points of a thick lens: (a) Focal (F$_1$, F$_2$) and principal points (H$_1$, H$_2$); (b) nodal points (N$_1$, N$_2$).

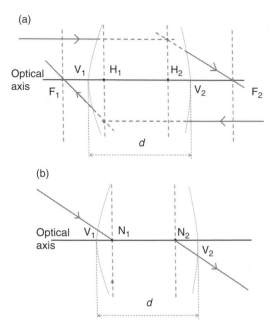

paraxial rays traveling parallel to the optical axes from right to left (from left to right) are converged at the first (second) focal point (Fig. 7.1a).

The definition of the principal points can be given as follows. Let us assume that each light ray of a bundle incident on a thick lens parallel to the axis is extended to meet the backward extension of the ray after passing through the system. The principal plane can be defined as the locus of the intersection of all the rays. By definition, the first (second) principal plane is formed by rays incident on the lens from the right (left).

Principal points are the intersections of the principal planes with the optical axis (Fig. 7.1a). There is also an equivalent definition: The principal points are a pair of points (one each in object space and image space) that are images of each other with magnification $M = +1$.

Nodal points are defined as two axial points located such that an oblique ray directed toward one point appears to emerge from the other parallel to its original direction (Fig. 7.1b). For a thick lens immersed in air, the nodal points coincide with the principal points ($H_i = N_i$, $i = 1,2$).

Figure 7.2 shows the positions of all six cardinal points of a thick lens for two cases: (a) When a thick lens borders media of different refractive indices (n_1 and n_2); (b) when a thick lens is immersed in a homogeneous medium of refractive index n_o.

The effective focal length f_{eff} is the distance from the second (first) principal point to the second (first) focal point (Fig. 7.2). Front (back) focal length is the distance from the front (back) vertex of the lens to the

(a)

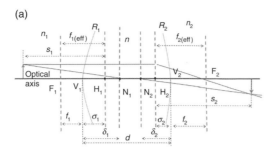

(b)

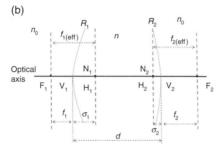

Figure 7.2 All six cardinal points of a thick lens: (a) Lens surfaces bordered with different media; (b) Lens is immersed in a homogeneous medium of refractive index n_o.

front (back) focal point. Front (back) vertex ($V_{1,2}$) is the intersection of the front (back) lens surface with the optical axis.

Let us summarize the basic equations for the thick lens. First, consider basic expressions for the most general case shown in Fig. 7.2a.

$$\frac{1}{f_{1(\text{eff})}} = \frac{n - n_2}{n_1 R_2} - \frac{n - n_1}{n_1 R_1} - \frac{(n - n_1)(n - n_2)}{n_1 n} \frac{d}{R_1 R_2} \tag{7.1}$$

$$f_{2(\text{eff})} = -\frac{n_2}{n_1} f_{1(\text{eff})} \tag{7.2}$$

$$\sigma_1 = \frac{n - n_2}{n R_2} f_{1(\text{eff})} d \tag{7.3}$$

$$\sigma_2 = -\frac{n - n_1}{n R_1} f_{2(\text{eff})} d \tag{7.4}$$

$$\delta_1 = \left(1 - \frac{n_2}{n_1} + \frac{n - n_2}{n R_2} d\right) f_{1(\text{eff})} \tag{7.5}$$

$$\delta_2 = \left(1 - \frac{n_1}{n_2} - \frac{n - n_1}{n R_1} d\right) f_{2(\text{eff})} \tag{7.6}$$

$$-\frac{f_{1(\text{eff})}}{s_1} + \frac{f_{2(\text{eff})}}{s_2} = 1 \tag{7.7}$$

◯ **Materials Needed**

- Optical table or optical rail
- Light sources (red cw laser)
- Object (arrow or cross)
- Ruler
- Mechanical holders
- Screws and screwdrivers
- Calipers and/or spherometer
- Squared graph paper
- Screen
- Needles
- Holding surface for mirror and prism
- Mirror
- Thick lenses: Plano-concave, plano-concave

In the case shown in Fig. 7.2b, $n_1 = n_2$, and the expressions simplify to:

$$\sigma_i = \delta_i, i = 1, 2 \qquad (7.8)$$

$$f_{\text{eff}} = f_{2(\text{eff})} = -f_{1(\text{eff})} \qquad (7.9)$$

$$\frac{1}{s_1} + \frac{1}{s_2} = \frac{1}{f_{\text{eff}}} \qquad (7.10)$$

All quantities in the equations above follow the usual sign conventions.

For thick lenses, the Newtonian lens formula is widely used to determine the location of all six cardinal points. For the thick lens shown in Fig. 7.2, the Newtonian lens formula can be written as follows:

$$\left(s_1 - f_{1(\text{eff})}\right)\left(s_2 - f_{2(\text{eff})}\right) = f_{\text{eff}}^2 \qquad (7.11)$$

Procedures

Wear gloves to handle optical elements (lenses).

Determination of All Six Cardinal Points of a Thick Lens

1. Identify thick lenses available in your optics kit.

2. Find a convex thick lens and assemble the experimental set-up shown in Fig. 7.3a.

3. Switch on the laser and move the mirror slowly in order to catch the auto-collimated image of the laser beam. Determine the first focal point F_1 of the thick lens. Additionally, measure the distance s between the nearby lens surface and the mirror. The measured distance is the first focal length f_1 of the thick lens.

   Note: For better results, expand the laser beam using a beam expander. Please refer to Chapter 8 for a description of how to build a simple beam expander made of two lenses.

4. Repeat the same procedures for the second surface of the thick lens as shown in Fig. 7.3b. To achieve better results, do not move the lens, but change the positions of the laser and mirror. Find the second focal point F_2 and measure the second focal length f_2.

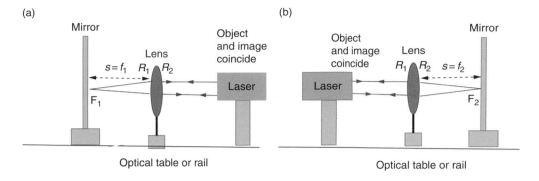

Figure 7.3 **Experimental set-up to find focal points of a thick lens.**

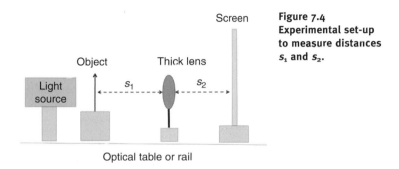

Screen

Figure 7.4
Experimental set-up to measure distances s_1 and s_2.

5. Put an object (arrow or cross) at a distance $s_1 > f_1$ and move the screen until the image is in focus on it, as shown in Fig. 7.4. Measure distances s_1 and s_2.

   Note: The distance s_1 (s_2) shown in Fig. 7.4 is measured between the object (image) and the front (back) vertex of the lens. Do not confuse these with distances s_1 (s_2) shown in Fig. 7.2.

6. Change distance s_1 and measure corresponding distance s_2. Complete the following table.

Table 7.1

Type of thick lens	First focal length f_1 (mm)	Second focal length, f_2 (mm)	Distance s_1 (mm)	Distance s_2 (mm)	$x_1 = s_1 - f_1$ (mm)	$x_2 = s_2 - f_2$ (mm)	f_{eff} * (mm)

* $x_1 x_2 = f_{eff}^2 \Rightarrow f_{eff} = \sqrt{x_1 x_2}$

7. Calculate the location of the principal and nodal points, and complete the following table.

Table 7.2

First focal length f_1 (mm)	Second focal length f_2 (mm)	f_{eff} (mm)	$\sigma_1 = f_{eff} - f_1$ (mm)	$\sigma_2 = f_{eff} - f_2$ (mm)

8. Draw a picture of a thick lens. Show all six cardinal points and measured distances.

Determination of the Refractive Index of a Thick Lens

1. Measure the thickness and aperture of a thick plano-convex lens using calipers.

2. Calculate the radius of curvature of a thick lens using the following expression (for details see Fig. 7.5):
   (a) Measure the lens thickness $d = BD$

   (b) Measure the lens aperture $D = AC$

   (c) Calculate the lens radius $R = OB = OC = OD + DB$:

$$R^2 = (R-d)^2 + \left(\frac{D}{2}\right)^2 \Rightarrow R = \frac{d^2 + \left(\frac{D}{2}\right)^2}{2d}$$

3. Calculate the refractive index of a thick lens by using the experimentally measured value of the effective focal length and the following expression ($n_1 = n_2 = 1$):

$$\frac{1}{f_{1(eff)}} = \frac{n - n_2}{n_1 R_2} - \frac{n - n_1}{n_1 R_1} - \frac{(n - n_1)(n - n_2)}{n_1 n} \frac{d}{R_1 R_2}$$

Figure 7.5 How to calculate the radius of a curvature of a lens. Consider the portion of the circle ABCDA as a cross-section of the thick lens.

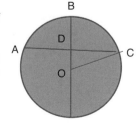

Note: For the case of a plano-convex lens, one of the radii R_1 or R_2 equals the value of R you just found, while the other radius equals infinity.

4. Complete the following table.

Table 7.3

d (mm)	D (mm)	R_1 (mm)	R_2 (mm)	f_{eff} (mm)	n

5. Draw a conclusion about the lens material.

Determination of All Six Cardinal Points of a Compound Thick Lens

Make a compound concave–convex thick lens with two thick lenses, plano-concave and plano-convex, in close contact. Use the procedure described above to determine all six cardinal points of the thick lens.

Evaluation and Review Questions

1. Derive equations for a thick lens.
2. Write equations for a thick lens immersed in a medium of refractive index n.
3. Draw all the cardinal points and planes of the studied thick lenses.

For Further Investigation

Read about the method called "the nodal slide" used to locate the cardinal points of a thick lens or a lens system. In this method, the location of the cardinal points is found using the fact that the image appears stationary if the lens (or a lens system) is rotated about the secondary nodal point N_2 (Figs. 7.1 and 7.2). A special mechanical holder for the lens, called the nodal slide, is needed. This holder allows both axial (horizontal) translation (along the optical axis) and rotation of the lens about the vertical axis. A typical experimental set-up is composed of the collimated light source and an object (in the majority of commercially available collimators, an object located at the focus of the collimator is a "built-in" feature of the device), a nodal slide with the lens (or lens system) under test, and a microscope to observe the image (the microscope can also be replaced with a screen, but such a replacement can result in less precision). The first step involves finding the focus of the lens.

The image of the object is in sharp focus on the screen or in the microscope. At this stage it is important to fix the distance between the lens and the screen (or the microscope) to ensure proper imaging of the object. After that, the nodal slide is used to rotate the lens about the vertical axis. The design of the nodal slide allows performing such a rotation of the lens about different vertical axes while keeping the image in focus. The goal of this manipulation is to achieve no lateral translation of the image while the lens is being rotated about the vertical axis. Typically, this is an iterative procedure involving manipulations with the nodal slide to minimize (ideally totally eliminate) lateral translation of the image. Once this is achieved, the axis of the lens rotations coincides with its rear nodal point. A nodal slide may be either purchased or self-made. Nodal slides are commercially available in a variety of styles because they are widely used by photographers interested in creating accurate stitched panoramic images.

For more details, please refer to the influential books on experimental optics listed below.

C. H. Palmer, *Optics: Experiments and Demonstrations*, Baltimore, MD: Johns Hopkins Press, 1962.

J. M. Geary, *Introduction to Optical Testing*, Bellingham, WA: SPIE Optical Engineering Press, 1993.

W. J. Smith, *Modern Optical Engineering*, 3rd edition, New York: McGraw Hill, 2000.

Further Reading

General

E. Hecht, *Optics*, 4th edition, San Francisco, CA: Addison-Wesley, 2001

F. L. Pedrotti, S. J. L. Pedrotti, L. M. Pedrotti, *Introduction to Optics*, 3rd edition, Upper Saddle River, NJ: Pearson Prentice Hall, 2007

J. Strong, *Concepts of Classical Optics*, New York: Dover, 2004 (Dover edition is an unabridged republication of the work originally published in 1958 by W. H. Freeman and Company, San Francisco, CA)

Specialized

Handbook of Optics, W. G. Driscoll (editor), W. Vaughan (associate editor), New York: McGraw-Hill, 1978

D. C. O'Shea, *Elements of Modern Optical Design*, New York: Wiley, 1985

W. J. Smith, *Modern Optical Engineering*, 3rd edition, New York: McGraw Hill, 2000

A. F. Wagner, *Experimental optics*, New York: John Wiley & Sons, 1929

Chapters 2, 5, and 6 of this book

Catalogs and Web Resources

www.cvimellesgriot.com - CVI Melles Griot
www.edmundoptics.com/ - Edmund Optics
www.newport.com/ - Newport Corporation
www.thorlabs.com/ - Thorlabs

8 Lens Systems

Objectives

1. Develop basic skills to manipulate two-lens systems.
2. Determine the locations of six cardinal points and planes of two-lens systems.
3. Formulate the limitations of the studied experimental techniques.

Background

Any complicated imaging optical system is a combination of simple optical components: Mirrors, lenses, prisms, plane plates, etc. Such optical systems are described in terms of Gaussian (or first-order) optics. In general, the approach using cardinal points and cardinal planes which we used to describe thick lenses (for details see Chapter 7 of this book) is also valid for more complicated optical systems. Definitions of all cardinal points and planes are the same as for thick lenses, so we refer the reader to the previous chapter for more details. Briefly, the basic idea of the description of an optical system in terms of cardinal points is as follows: *six cardinal points and corresponding cardinal planes can replace an optical system.* If you know all these cardinal elements, you can make a ray tracing and build an image of any object (of course, in the paraxial approximation).

Let us consider an optical system immersed in air, as shown in Fig. 8.1. In this case, nodal (N_i, $i = 1$, 2) and principal (H_i, $i = 1$, 2) points coincide. Additionally, the front and back effective focal lengths are equal, so the term "effective focal length" f_{eff} is used instead of the two terms "front and back effective focal lengths."

The image position is found by the following useful and simple expressions:

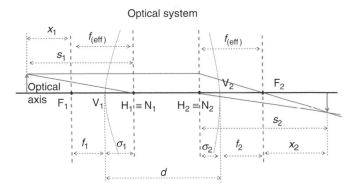

$$\frac{1}{s_1} + \frac{1}{s_2} = \frac{1}{f_{\text{eff}}} \tag{8.1}$$

$$x_1 x_2 = f_{\text{eff}}^2 \tag{8.2}$$

where x_1 is the distance between the object and front (first) focal point F_1, and x_2 is the distance between the image and the back (second) focal point F_2 (Fig. 8.1). Distances s_1 (s_2) are measured between the object (image) and the first (second) principal plane. As seen here, these equations are similar to the equations describing a thin lens, and this fact can help one to understand the basic idea of representing the optical system by cardinal elements.

The combination of two thin lenses is a simple but very useful example of optical systems. Since two-lens combinations can be found in many applications, it is important to describe them in more detail.

Align two thin converging lenses along the optical axis and set a distance d between them as shown in Fig. 8.2. Lens 1 has focal length $f^{(1)}$, and front (back) focal points $F^{(1)}_1$ ($F^{(1)}_2$). Analogously, lens 2 has focal length $f^{(2)}$, and front (back) focal points $F^{(2)}_1$ ($F^{(2)}_2$). Cardinal elements of such optical systems can be determined according to the following general procedure. Draw an incident light ray, propagating from the left to the right, in parallel to the optical axis. Using the standard geometrical technique of ray tracing for thin lenses (details can be found, for example, in Chapter 6), find the back focal point F_2 of the optical system. This is the intersection of the output light ray and the optical axis. Extend the incident light ray to meet the light ray (or its backward extension) after passing through the system. Intersection of these rays defines the back principal plane, which intersects the optical axis in back principal point H_2. Analogously, by considering a light ray traveling from the right to the left, front focal point F_1 and front principal point H_1 can be found, as shown in Fig. 8.2. The distance H_1F_1 equals distance H_2F_2: $H_1F_1 = H_2F_2$. This distance is called the effective focal length f_{eff} of the two-lens

Figure 8.2 Combinations of two thin lenses: (a) The distance d between the lenses is greater than the sum of their focal lengths, $d > f^{(1)} + f^{(2)}$; (b) the distance d between the lenses is smaller than the sum of their focal lengths, $d < f^{(1)} + f^{(2)}$.

Note: Work on these two images diligently. The ray tracing here is exact; make sure you understand these principles.

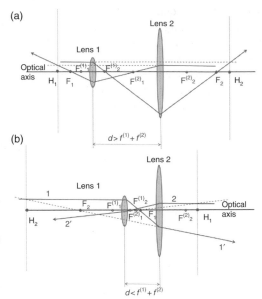

Figure 8.3 Ray tracing of an object (arrow) placed in front of a two-lens system: (a) Step-by-step ray tracing; (b) ray tracing using the formalism of cardinal elements.

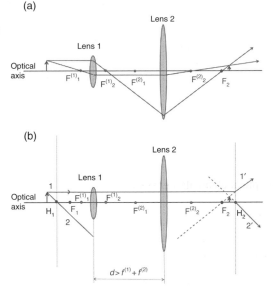

system. Since the lenses are immersed in air, principal points $H_{1,2}$ coincide with the nodal points $N_{1,2}$: $H_{1,2} \equiv N_{1,2}$.

Figure 8.3 shows how to draw the image of an object for the two-lens system considered above using two methods. First, use simple ray tracing where each lens is considered separately. The image due to the first lens is the object for the second lens. Second, use cardinal elements of the optical system. Knowing the cardinal elements of an optical system simplifies ray tracing and saves time, especially when the optical system is composed of many lenses.

The effective focal length f_{eff}, and the front f_1 and back f_2 focal lengths of a two-component system (for instance, two thin lenses with focal lengths $f^{(1)}$ and $f^{(2)}$, separated by a distance d) can be calculated from the following expressions:

$$\frac{1}{f_{eff}} = \frac{1}{f^{(1)}} + \frac{1}{f^{(2)}} - \frac{d}{f^{(1)}f^{(2)}} \qquad (8.3)$$

$$f_{eff} = \frac{f^{(1)}f^{(2)}}{f^{(1)} + f^{(2)} - d} \qquad (8.4)$$

$$f_1 = \frac{f_{eff}\left(f^{(2)} - d\right)}{f^{(2)}} \qquad (8.5)$$

$$f_2 = \frac{f_{eff}\left(f^{(1)} - d\right)}{f^{(1)}} \qquad (8.6)$$

Since the front and back focal lengths of the optical system and the effective focal length can be measured experimentally, the locations of the principal and nodal points can be determined.

Materials Needed

- Optical table or optical rail
- Light sources (LED, red cw laser, tungsten bulb)
- Voltage supply
- 10 Ω resistor
- Object (arrow or cross)
- Ruler
- Mechanical holders
- Screws and screwdrivers
- Calipers and/or spherometer
- Squared graph paper
- Screen
- Needles
- Mirror
- Thin lenses: Plano-concave, plano-concave

Procedures

Wear gloves to handle optical elements (lenses).

Making a Collimated Light Source

1. Identify converging lenses available in your optics kit and measure their aperture diameters D using calipers.

2. Estimate the focal length of converging lenses by using the following express technique – "object at infinity." Assemble the set-up shown in Fig. 8.4 and find the focused image on the screen of any object located at infinity. This means that the object distance should be much greater than the focal length of the lens. For example, use entrance doors, windows, or illuminating bulbs as "the object at infinity." According to the

**Figure 8.4
Determination of the
focal length of a
converging lens using
the "object at
infinity" method.**

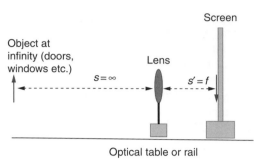

Object at
infinity (doors,
windows etc.)

Screen

Lens

$s = \infty$

$s' = f$

Optical table or rail

**Figure 8.5 Electrical
circuit: Light source
(incandescent lamp),
resistor (10–15 Ω),
and voltage supply
(5 V) connected in
series.**

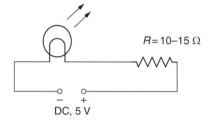

$R = 10\text{–}15\ \Omega$

DC, 5 V

lens equation $\frac{1}{s} + \frac{1}{s'} = \frac{1}{f}$, the distance between the lens and the screen s' is the focal length of the lens, $f = s'$ (because $s = \infty$).

Note: The next experiment is described using a tungsten lamp. You may instead want to consider using an LED as it works better as a point light source.

3. Identify a tungsten bulb available in your optics kit and design the electrical circuit shown in Fig. 8.5.

4. Assemble the set-up shown in Fig. 8.6. Move the lens along the optical rail and simultaneously observe the image of the tungsten bulb on the screen. Make the distance s between the bulb and the lens to equal the focal length f of the lens used. Use the screen to check the correct position of the lens. Adjust the lens so that the light passing through the lens forms a parallel light beam. This means that if you change distance s' between the lens and the screen, the diameter d of the light beam on the screen does not change. This is only an approximation because a real "parallel" light beam has a finite divergence angle. For the screen position found, measure the distance s'_1 between the lens and the screen and the diameter d_1 of the collimated light beam on the screen (see Fig. 8.6 for details).

5. Place the screen at the furthest distance you can. Measure the distance s'_2 between the lens and the screen, and the diameter

Lens Systems

(a)

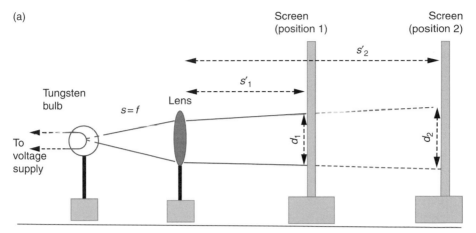

Optical table or rail

(b)

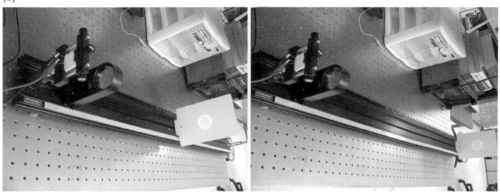

d_2 as shown in Fig. 8.6. Calculate the divergence angle of your collimated light beam using the following expression (Fig. 8.7 explains the meaning of the angle α): $\tan \frac{\alpha}{2} = \frac{1}{2} \frac{d_2 - d_1}{s'_2 - s'_1}$.

Figure 8.6
Experimental set-up to produce a collimated light source.

6. Repeat all the above-listed measurements for converging lenses of different aperture diameters and focal lengths. Complete the following table.

Table 8.1

Converging lens	Lens aperture diameter, D (mm)	Focal length, f (mm)	s'_1 (mm)	d_1 (mm)	s'_2 (mm)	d_2 (mm)	tan $(\alpha/2)$	α (deg/rad)
#1								
#2								
#3								

Figure 8.7 Divergence angle of a quasi-collimated light beam.

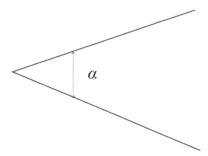

Figure 8.8 Optical set-up of laser beam expander.

(a)

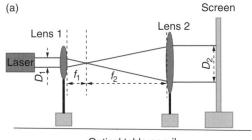

(b)

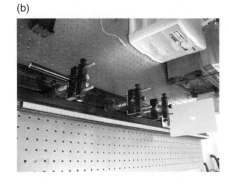

7. Propose ways to improve the performance of the collimated light source.

Laser Beam Expander

1. Identify two converging lenses in your optics kit according to the following descriptions.

   Lens 1: Focal length 18 mm, aperture diameter 12 mm

   Lens 2: Focal length 100 mm, aperture diameter 25 mm

2. Assemble the optical set-up shown in Fig. 8.8.

3. Fix lens 1, and move lens 2 slowly until the back focal point of lens 1 and front focal point of lens 2 coincide (Fig. 8.8). Observe the expanded laser beam on the screen.

4. Measure the diameter D_2 of the expanded laser beam. Calculate the diameter D_1 of the incident laser beam using the following expression: $\frac{D_2}{D_1} = \frac{f_2}{f_1}$, where f_i is the focal length of the ith lens ($i = 1, 2$). Complete the following table.

Table 8.2

Lens	Focal length, f (mm)	Lens aperture diameter, d (mm)	Diameter of expanded laser beam, D_2 (mm)	Diameter of incident laser beam, D_1 (mm)	D_2/D_1
#1					
#2					

5. Discuss factors affecting the performance of the laser beam expander.

Cardinal Points of a Combination of Two Thin Converging Lenses, 1

1. Using the two converging thin lenses found in the previous procedure, arrange a two-component optical system. Place two lenses along the optical axis separated by a distance $d > f^{(1)} + f^{(2)} = 12$ cm, as shown in Fig. 8.9.

2. Switch on the laser and move the mirror slowly to catch the auto-collimated image of the laser beam. This means that if the mirror is adjusted properly, the laser beam reflected after passing through the optical system coincides with the incident laser beam at the laser output. Determine the second focal point F_2 of a two-component optical system. Additionally, measure the distance between lens 2 and the mirror. The measured distance is the second (or back) focal length f_2 of the two-lens system (Fig. 8.9a).

Figure 8.9
Experimental optical set-up: Auto-collimating method to determine focal points of a two-component optical system.

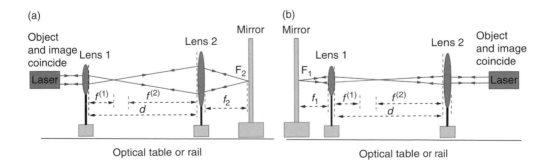

Optical table or rail   Optical table or rail

Figure 8.10
Experimental set-up
to measure distances
x_1 **and** x_2.

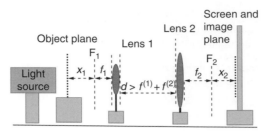

Optical table or rail

3. Repeat the same procedure, as shown in Fig. 8.9b. To achieve better results, do not move the lenses, but change the positions of the laser and mirror. Find first focal point F_1 and measure the first (front) focal length f_1.

4. Put an object (an arrow or a cross) at a distance $x_1 > f_1$ and move the screen in order to bring the image into focus on the screen, as shown in Fig. 8.10. Measure distances x_1 and x_2.

5. Change the distance x_1 and measure the corresponding distance x_2. Complete the following table.

Table 8.3

Measured quantities							Calculated quantities	
Distance between the lenses, d (mm)	First focal length, f_1 (mm)	Second focal length, f_2 (mm)	Distance x_1 (mm)	Distance x_2 (mm)	f_{eff}* (mm)	f_{eff} (mm)	First focal length, f_1 (mm)	Second focal length, f_2 (mm)

* $x_1 x_2 = f_{eff}^2 \Rightarrow f_{eff} = \sqrt{x_1 x_2}$

6. Determine the location of the principal and nodal points. Use Fig. 8.2 as a guide.

7. Draw a picture using an appropriate scale, and show all six cardinal points, indicating measured distances.

Cardinal Points of a Combination of Two Thin Converging Lenses, 2

Put lenses 1 and 2 along the optical axis separated by a distance $d < 10$ cm, as in Fig. 8.9. Repeat the procedures described in the previous section, and find all cardinal elements. Complete the following table.

Table 8.4

Measured quantities				Calculated quantities		
Distance between the lenses, d (mm)	First focal length, f_1 (mm)	Second focal length, f_2 (mm)	f_{eff} (mm)	f_{eff} (mm)	First focal length, f_1 (mm)	Second focal length, f_2 (mm)

Hint: This can be confusing because the focal points are virtual. Use the approach to find focal points of the diverging lenses as described in Chapter 6.

Cardinal Points of a Combination of Converging and Diverging Lenses

Put converging and diverging lenses along the optical axis separated by a distance $d < 10$ cm. Find all the cardinal elements of such a system using geometrical ray tracing. Draw all the cardinal points and planes of this combination.

Evaluation and Review Questions

1. Derive all the equations shown in the background section.
2. Choosing an appropriate scale, draw all the cardinal points and planes of the studied optical systems of two thin lenses.

For Further Investigation

1. Estimate the "efficiency" of your collimating system. The "efficiency" is the ratio of the collimated light beam luminance to the total output of luminance of the tungsten bulb. Hint: Consider a tungsten bulb as a point light source, and use the inverse square law (studied diligently by yourself in Chapter 4).

2. Make and study a so-called telephoto lens. This is a combination of converging and diverging lenses immersed in air and separated by such a distance as to obtain a comparatively long effective focal length.

3. Draw the aperture stop, entrance and exit pupils, and entrance and exit windows for the laser beam expander shown in Fig. 8.8.

Further Reading

General

F. L. Pedrotti, S. J. L. Pedrotti, L. M. Pedrotti, *Introduction to Optics,* 3rd edition, Upper Saddle River, NJ: Pearson Prentice Hall, 2007

E. Hecht, *Optics*, 4th edition, San Francisco, CA: Addison-Wesley, 2001

J. Strong, *Concepts of Classical Optics*, New York: Dover, 2004 (Dover edition is an unabridged republication of the work originally published in 1958 by W. H. Freeman and Company, San Francisco, CA)

Specialized

Handbook of Optics, W. G. Driscoll (editor), W. Vaughan (associate editor), New York: McGraw-Hill, 1978

D. C. O'Shea, *Elements of Modern Optical Design*, New York: Wiley, 1985

W. J. Smith, *Modern Optical Engineering*, 3rd edition, New York: McGraw Hill, 2000

A. F. Wagner, *Experimental Optics*, New York: John Wiley & Sons, 1929

Chapter 7 of this book

Simple Optical Instruments I: The Eye and the Magnifier, Eyepieces, and Telescopes

9

Objectives

1. Develop basic skills to make and manipulate simple optical instruments: Magnifier, telescope, and eyepiece.
2. Determine the locations of aperture stops, field stops, entrance and exit pupils, and entrance and exit windows of the systems.
3. Formulate general practical rules about how the performance of these optical instruments influences their image-making properties.

Background

Real optical systems are always limited in size, so it is important to know how the size of an optical system affects its performance. Consider the simplest optical system composed of one thin lens, as shown in Fig. 9.1. As we increase the angle between the incident ray emitted from the on-axis point O and the optical axis, the ray will eventually hit the edge of the lens. For angles beyond this slope angle, the light rays will not be imaged by the lens. An axial ray traced near the outer edge of the system (e.g. the lens rim) is termed a *marginal ray*. The physical opening in an optical system that limits the amount of light that can be collected is called the *aperture stop* of the system (the lens rim serves as the aperture stop in Fig. 9.1). A *chief ray* is a ray propagating through the center of the aperture stop.

An aperture stop is a very important concept in optical system design because it determines the ability of the optical system to collect light. In other words, the aperture stop limits the brightness of the image. The image of the aperture stop, as seen from the object space (the space

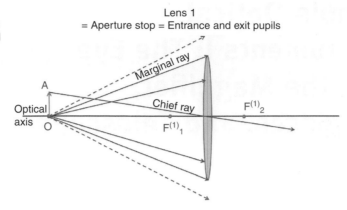

to the left of the first element of the optical system), is called the *entrance pupil* of the system. The image of the aperture stop as seen from the image space is called the *exit pupil* of the system. In our simple example of an aperture stop, entrance and exit pupils coincide (Fig. 9.1).

There is a simple algorithm (described below) to find the entrance pupil, the aperture stop, and the exit pupil.

(1) Make images of all the optical elements of the optical system as seen from the object space.

(2) Find the angle subtended by each element image at the on-axis position of the object.

(3) The element image with the smallest angle is the entrance pupil.

(4) The physical object corresponding to this image is the aperture stop.

(5) Make an image of the aperture stop as seen from the image space. This image is the exit pupil.

A practical realization of this algorithm is shown in Fig. 9.2 for an optical system composed of two thin lenses and an aperture placed between them. Figure 9.2 demonstrates that the rim of lens 2 is the aperture stop of the optical system despite it having the largest aperture diameter. We suggest the use of grid paper to make exact drawings and practice finding these elements on your own.

Also, we need to know what limits the extent of the image. It is obvious that we can change the slope of the chief ray until a certain maximal value; the slope of the chief ray can be increased until it is cut off by the edge of a component, as shown in Fig. 9.3a (dashed chief ray). This component, which limits the chief ray, is the *field stop* of the system. The

(a)

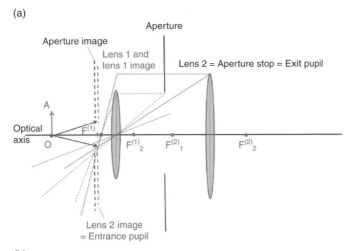

(b)

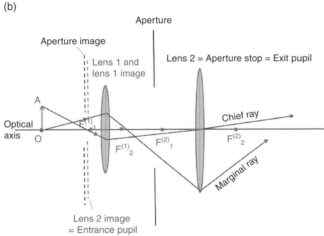

Figure 9.2 The optical system, composed of two thin lenses and an aperture ($F^{(1)}_1$ and $F^{(1)}_2$ are the front and back focal points of lens 1, $F^{(2)}_1$ and $F^{(2)}_2$, the front and back focal points of lens 2. O is the on-axis object point. (a) Aperture stop, entrance and exit pupils of the system. (b) Chief and marginal rays of the system. Note: All optical systems for visual observations by eye are designed in such a way that the exit pupil of the system approximately coincides with the pupil of the observer's eye (hence the origin of the term "pupil" can be understood).

image of the field stop formed by the optical system as seen from the object space (before the field stop) is termed the *entrance window*. When the field stop is in the object space, the field stop and entrance window coincide. The image of the field stop formed by the optical system as seen from the image space (after the field stop) is termed the *exit window*. The angle subtended by the marginal chief rays at the entrance pupil is the *angular field of view* of the system. It should be highlighted that the exit window is not in the same plane as the image of the object. This means that the edges of the image field will be blurred and out of focus. The planes of the image and the exit window will coincide only if the field stop is at the object, at an intermediate image, or at the final image.

There is also a simple algorithm (described below) to find the entrance window, the field stop, and the exit window.

Figure 9.3 An optical system composed of two thin lenses and an aperture: Aperture stop, entrance and exit pupils, field stop, and entrance and exit windows of the system.

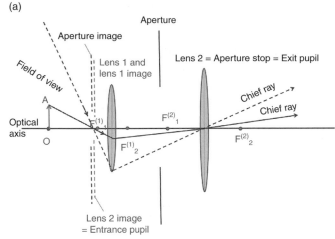

(a)

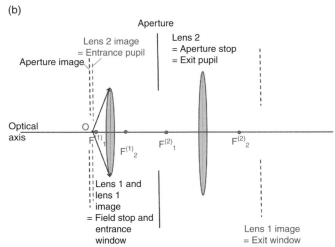

(b)

(1) Make images of all the optical elements of the optical system as seen from the object space.

(2) Find the angle subtended by each element image at the on-axis position of the entrance pupil.

(3) The element image with the smallest angle is the entrance window.

(4) The physical object corresponding to this image is the field stop.

(5) Make an image of the field stop as seen from the image stop. This image is the exit window.

Figure 9.3b shows a practical realization of the algorithm described above.

There are several key differences between an aperture stop and a field stop:

- The aperture stop determines the solid angle of the transmitted light cone for an *on-axis* object and limits the brightness of the image.

- The field stop determines the solid angle formed by the chief rays from the *off-axis* objects and limits the field of view of an optical instrument.

A cone of light rays from an off-axis object to the entrance pupil will not necessarily be transmitted in its entirety. It can be partially cut off by field stops or lens rims in the system. As a result, the brightness of the image is weaker at its edges than at its center. This phenomenon is called *vignetting*.

The concepts of aperture/field stop and entrance and exit pupil/ windows are vitally important in the analysis of an optical system. The placement and sizes of these quantities have a profound effect not only on the light-collecting ability of the optical system, but also on certain aberrations (more details on aberrations can be found in Chapter 12 of this book).

Simple Optical Instruments

The Eye

The human eye is a quasi-spherical, jelly-like object (the average size of its cross-section is ~24 mm × 22 mm) located inside a tough, flexible shell called the sclera. The sclera is white and opaque except for its front portion called the cornea, which is transparent. The eye is a natural optical device composed of multiple refractive elements (the cornea, aqueous humor, iris, and lens) and a light-detecting element (the retina). A schematic cross-section of the human eye is shown in Fig. 9.4.

Light enters the eye through the transparent front surface called the cornea. A watery fluid, the aqueous humor, fills the space between the cornea and the iris. The iris controls the amount of light entering the lens, which lies behind the iris. Depending on the intensity of the light entering the eye, the diameter of the hole in the iris, called the pupil, can vary from approximately 2 mm (very bright light) to 8 mm (very dim light). Thus, the iris serves as the aperture stop. It also gives the eye its characteristic color.

Light refracted by the cornea and passed through the iris is focused by the lens onto the retina. The space between the lens and

Figure 9.4
(a) Schematic representation of the human eye from an optical perspective.
(b) The cone spacing determines the minimum angular resolution of the eye

(redrawn after R. Ditteon, *Modern Geometrical Optics*).

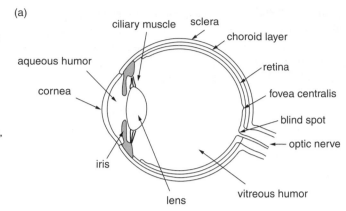

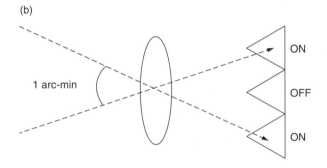

retina is filled with a transparent, gelatinous substance called the vitreous humor.

The retina is the image plane of the eye (in other words, the image of an object forms on the retina). Rods and cones are the light-sensing cells (or light receptors) of the retina and are named thus because of their shape. The rods can detect very low levels of light, but they do not sense color. The cones are sensitive to different wavelengths of light (color vision, approximately 390–780 nm wavelength), but they require more light to function. The distribution of the surface concentration of rods and cones is very inhomogeneous. There is a region with no light receptors which is insensitive to light and known as the blind spot (this is the area of exit of the optic nerve from the eye). Approximately at the center of the retina, a small yellow spot (2.5–3 mm in diameter) called the macula is located. The central region of the macula (~0.2 mm in diameter) is rod free and is occupied by densely packed cones. This region is called the fovea centralis or fovea. It provides the sharpest and most detailed information on objects observed by the eye. To achieve better vision, the eye moves constantly so the observed objects are imaged on the fovea.

The capability of the human eye to resolve two closely-spaced objects depends on the angular spacing of the light-sensing cells (rods and cones). The angular spacing between the neighboring cones in the fovea is approximately 30 arc-seconds. To resolve two objects, light must excite two cones separated by at least one cone which is not illuminated by the light. As a result, the eye can resolve two objects separated by at least 2×30 arc-seconds, equal to 1 arc-minute, as shown in Fig. 9.4b.

The distance between the cornea and the retina (~24 mm) is fixed. To see objects located at different distances from the eye, the shape of the lens is changed by means of the surrounding muscles. Thus, the lens becomes thicker to view nearby objects, and thinner to focus on distant objects. The property of the human eye to view both distant and nearby objects (by changing the shape of the lens) is called accommodation. The greatest distance at which the human eye can focus is called the far point. Ideally, it should be at infinity. The distance to the closest object the eye can still see in sharp focus is known as the near point. The standard value of the near point is 25 cm. Many optical instruments are designed in such a way that the image they form is located at infinity. Thus, the eye is in a relaxed state while observing the image.

The Magnifier

A simple converging lens can be used as a magnifier when an object's distance is smaller than the lens's focal length. In this case, we observe a magnified virtual image. Magnification M can be defined as the ratio of image height H to object height h. It is difficult to use this definition for objects or images at infinity, therefore a more appropriate definition for the magnification can be formulated:

$$M = \frac{H}{h} = \frac{\tan{(\alpha_M)}}{\tan{(\alpha_o)}} \approx \frac{\alpha_M}{\alpha_o} \tag{9.1}$$

The meaning of angles α_o and α_M can be seen from Fig. 9.5.

There is a simple relation between the magnification and the focal length of a lens:

$$M = \frac{250}{f} + 1; \text{image at normal near point } (S = 250 \text{ mm}) \tag{9.2}$$

$$M = \frac{250}{f}; \text{image at infinity } (S = \infty) \tag{9.3}$$

Eyepieces

An eyepiece is a combination of two lenses separated by a certain distance $d < f^{(1)} + f^{(2)}$, where $f^{(i)}$ is the focal length of the ith lens ($i = 1, 2$). Figure 9.7 (in the Procedures section) shows the structure of an eyepiece schematically. An eyepiece is the output part of an optical

Figure 9.5 Angular view of the object (a) and magnified image (b).

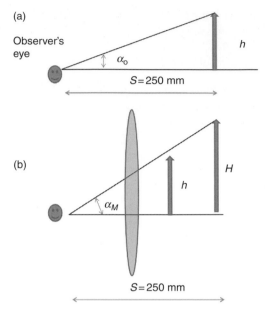

(a)

Observer's eye

α_o

h

$S = 250$ mm

(b)

α_M

h

H

$S = 250$ mm

system; it views not an actual object, but the intermediate image of the object as formed by the preceding lens system. The preceding lens system forms a so-called intermediate image, and then the eyepiece forms a virtual image of the intermediate image, most often located at or near infinity. As a result, the image formed by the eyepiece can be viewed by the normal, relaxed eye.

There are different types of eyepieces, such as the Huygens, Ramsden, Kellner, orthoscopic, symmetrical, and Erfle eyepieces.

The performance of eyepieces is determined by:

(1) Angular magnification (as a rule, angular magnification is between $4\times$ and $25\times$);

(2) Eye relief, or the distance between eye lens and exit pupil;

(3) Field of view or size of the primary image that the eyepiece can cover (varies from 6 to 30 mm).

Refracting Telescopes

The simplest astronomical telescope consists of two convex lenses of different focal lengths $f^{(1)}$ and $f^{(2)}$, which are aligned along the optical axis (Procedures section, Fig. 9.9). The first lens, on which the light falls first, is called the object lens. The second is known as the eye lens. The focal length of the object lens $f^{(1)}$ is much greater than the focal length $f^{(2)}$ of the eye lens: $f^{(1)} \gg f^{(2)}$. Also, the back focal point of the object lens coincides with the front focal point of the eye lens. The ratio $f^{(1)}/f^{(2)}$ determines the magnification of the telescope: $M = f^{(1)}/f^{(2)}$.

Telescopes are designed so that the object lens serves as an aperture stop and an entrance pupil, while the eye lens serves as a field stop.

Modern astronomical telescopes have more complicated object and eye lenses. Different achromatic lenses are used as object lens and eyepieces are used as the eye lens.

> **Materials Needed**
>
> - Optical table or optical rail
> - Light sources (red cw laser, tungsten bulb)
> - Object (arrow or cross)
> - Ruler
> - Mechanical holders
> - Screws and screwdrivers
> - Calipers and/or spherometer
> - Squared graph paper
> - Screen
> - Needles
> - Mirror
> - Optics kit including thin lenses: Plano-concave, plano-convex

Procedures

Wear gloves to handle optical elements (lenses).

Converging Lens as Simple Magnifier

1. Identify converging lenses available in your optics kit and measure their aperture diameters D using calipers.

2. Estimate the focal length f of the converging lenses using any suitable technique you studied in previous procedures. For all converging lenses, calculate a theoretically possible value of their angular magnification, M_{theory}, assuming that the image is located both at a normal near point ($M_{theory,25}$) and at infinity ($M_{theory,\infty}$).

3. Place a lens in front of a piece of squared paper at a distance d which is smaller than the lens focal length f: $d < f$. Observe the paper through the lens and, by using a ruler, measure the period H of the magnified square grid. Additionally, measure the period h of the same square grid using the unaided eye (without the lens, Fig. 9.6). For chosen distance d, calculate the real angular magnification $M_{experiment}$ of the lens using the experimentally measured values H, h.

4. Change distance d and repeat the same measurements. Complete the following table.

Table 9.1

Converging lens	Lens aperture diameter, D (mm)	Focal length, f (mm)	$M_{\text{theory},25}$ *	$M_{\text{theory},\infty}$ **	d (mm)	h (mm)	H (mm)	$M_{\text{experiment}}$ ***
#1								
					$d = f$			
#2								
					$d = f$			
#3								
					$d = f$			

* $M_{\text{theory},25} = 1 + 250/f$
** $M_{\text{theory},\infty} = 250/f$
*** $M_{\text{experiment}} = H/h$

5. Analyze and explain the experimental results obtained.

Figure 9.6 (a) Determination of the magnification of a converging lens. (b) Magnified image of an object through a magnifier.

(a)

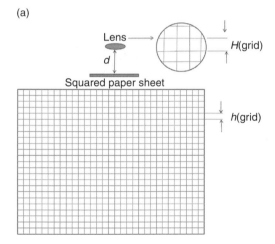

(b)

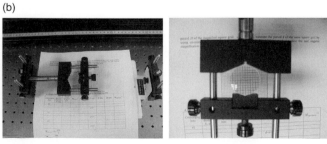

Ramsden Eyepiece

1. Identify two identical plano-convex lenses in your optics kit according to the following description.

   Lens parameters: Focal length f ~18 mm or ~20–25 mm, aperture diameter ~12 mm or ~25 mm.

2. Place two plano-convex lenses along the optical axis and fix a distance $d = \frac{2}{3}f$ between them, as shown in Fig. 9.7a.

3. Arrange the optical set-up shown in Fig. 9.7b. It is a good idea to expand the laser beam (see Chapter 8 for more details). Switch on the laser and move the mirror slowly to catch the auto-collimated image of the expanded laser beam. If the mirror is adjusted properly, the reflected laser beam, after passing through the optical system, coincides with the incident laser beam on the laser output. Determine the first focal point F_1 of the Ramsden eyepiece. Additionally, measure the distance between the mirror and the nearest lens. The measured distance is the first (or front) focal length f_1 of the Ramsden eyepiece (Fig. 9.7b).

(a)

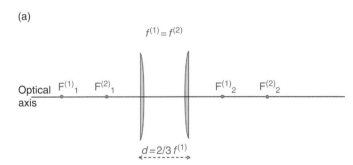

(b)

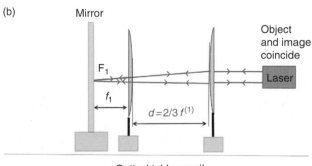

Figure 9.7   Ramsden eyepiece: (a) Optical scheme; (b) experimental set-up to measure focal points of the eyepiece.

Figure 9.8 (a) Experimental set-up to measure distances x_1 and x_2. (b) Seeing an object through a Ramsden eyepiece.

(a)

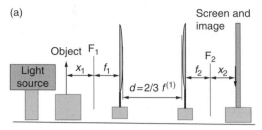

(b)

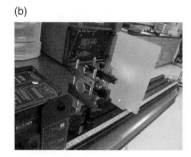

4. Repeat the same procedures as shown in Fig. 9.7b and find the second focal point F_2. Measure the second (back) focal length f_2.

5. Put an object (an arrow or a cross) at a distance $x_1 > f_1$ and move the screen to bring the image into focus on it, as shown in Fig. 9.8. Measure distances x_1 and x_2.

6. Change the distance x_1 and measure the corresponding distance x_2. Create and complete Table 9.2, summarizing these experimental data.

Table 9.2

Measured quantities							Calculated quantities		
Distance between the lenses, d (mm)	First focal length, f_1 (mm)	Second focal length, f_2 (mm)	Distance, x_1 (mm)	Distance, x_2 (mm)	f_{eff} * (mm)	f_{eff} (mm)	First focal length, f_1 (mm)	Second focal length, f_2 (mm)

$^* x_1 x_2 = f_{eff}^2 \Rightarrow f_{eff} = \sqrt{x_1 x_2}$

7. Determine the locations of the principal and nodal points.

8. Determine the location of the aperture stop, entrance pupil, exit pupil, field stop, entrance window, and exit window for the studied Ramsden eyepiece.

9. Draw a picture using an appropriate scale. Show all six cardinal points, aperture stop, entrance pupil, exit pupil, field stop, entrance window, and exit window. Indicate measured distances.

Refracting Astronomical Telescope

1. Identify two plano-convex lenses in your optics kit according to the following description.

   Lens 1: Focal length ~100 mm, aperture diameter ~25 mm

   Lens 2: Focal length ~18 mm, aperture diameter ~12 mm

2. Assemble the experimental set-up shown in Fig. 9.9.

3. Align all optical elements of the set-up shown in Fig. 9.9 along the optical axis. Put the light source with the object (two crossed arrows) at distance $s \gg d$ as shown in Fig. 9.9. Fix lens 1, and move lens 2 slowly; simultaneously observe the image of the object through lens 2 with your eye. Your eye needs to be very close to lens 2, almost touching it with your eyelashes. Adjust the position of lens 2 to bring the image of the object into focus on the retina of your eye.

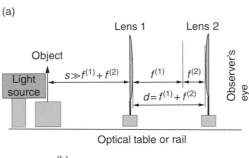

(a)

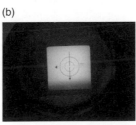

(b)

Figure 9.9
(a) Refracting astronomical telescope. (b) Seeing an object through a refracting astronomical telescope.

4. Note the type of image you see: erect or inverted?

5. Show graphically that lens 1 serves as an aperture stop and entrance pupil while lens 2 serves as a field stop. To do this, draw a picture of the refracting telescope, using appropriate scale, and show the aperture stop, entrance pupil, exit pupil, field stop, entrance window, and exit window. Indicate the measured distances.

6. Measure the sizes of the entrance and exit pupils, $D_{entrance}$ and D_{exit}. Since the entrance pupil and aperture stop coincide, its size can be measured directly by measuring the diameter of lens 1. To measure the size of the exit pupil, put a cardboard screen behind lens 2 and bring the output light beam into focus (the beam emerging from the lens). Next, measure the diameter of such an output light beam, or the size of the exit pupil.

7. Calculate the magnification of the studied telescope and complete the following table.

Table 9.3

Measured quantities						Calculated quantities
Distance between lenses 1 and 2, d (mm)	Lens 1 focal length, $f^{(1)}$ (mm)	Lens 2 focal length, $f^{(2)}$ (mm)	$D_{entrance}$ (mm)	D_{exit} (mm)	Magnification, $M = D_{entrance}/D_{exit}$	Magnification, $M = f^{(1)}/f^{(2)}$

Refracting Terrestrial Telescope

1. Identify three plano-convex lenses in your optics kit according to the following description:

   Lens 1: Focal length ~100 mm, aperture diameter ~25 mm

   Lens 2: Focal length ~18 mm, aperture diameter ~12 mm

   Lens 3: Focal length ~18 mm, aperture diameter ~12 mm

2. Assemble the experimental set-up shown in Fig. 9.10.

3. Align all the optical elements of the set-up shown in Fig. 9.10 along the optical axis. Put the light source with the object (two crossed arrows) at distance $s \gg d$ $(d = f^{(1)} + 4f^{(3)} + f^{(2)})$, as shown in Fig. 9.10.

4. Note the type of image you see: erect or inverted?

The Eye and the Magnifier, Eyepieces, and Telescopes

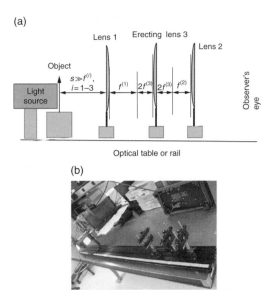

(a)

Figure 9.10
Refracting terrestrial telescope.

(b)

5. **Optional:** Draw a picture of the refracting telescope using an appropriate scale and show the aperture stop, entrance pupil, exit pupil, field stop, entrance window, and exit window. Indicate measured distances.

6. Measure the sizes of the entrance and exit pupils, $D_{entrance}$ and D_{exit}. Since the entrance pupil and aperture stop coincide, its size can be measured directly by measuring the diameter of lens 1. To measure the size of the exit pupil, put the cardboard screen behind lens 2 and bring into focus the output light beam (the beam emerging from the lens). Next, measure the diameter of such an output light beam, or the size of the exit pupil.

7. Calculate the magnification of the studied telescope and complete the table below.

Table 9.4

Measured quantities						Calculated quantities
Distance between lenses 1 and 2, d (mm)	Lens 1 focal length, $f^{(1)}$ (mm)	Lens 2 focal length, $f^{(2)}$ (mm)	$D_{entrance}$ (mm)	D_{exit} (mm)	Magnification, $M = D_{entrance}/D_{exit}$	Magnification, $M = f^{(1)}/f^{(2)}$

8. Explain the role of the erecting lens.

Refracting Telescope with an Eyepiece

Assemble the Ramsden eyepiece shown in Fig. 9.7 and replace lens 2 (Fig. 9.9) with a Ramsden eyepiece. Perform the same procedures described in the two previous sections for the telescope with eyepiece. Draw a conclusion about improvements in the performance of the telescope when using an eyepiece instead of the single eye lens (lens 2 in our case).

Evaluation and Review Questions

1. Derive all equations shown in the Background section.
2. Prove that magnification of the telescope can be defined as the ratio of the diameters of the entrance and exit pupils.
3. Choosing an appropriate scale, draw all the cardinal elements, aperture stops, field stops, pupils, and windows of the studied optical systems of two thin lenses.

For Further Investigation

Find optical schemes of modern eyepieces. Identify the aperture stops, field stops, pupils, and windows for them.

Further Reading

General

E. Hecht, *Optics*, 4th edition, San Francisco, CA: Addison-Wesley, 2001

F. L. Pedrotti, S. J. L. Pedrotti, L. M. Pedrotti, *Introduction to Optics*, 3rd edition, Upper Saddle River, NJ: Pearson Prentice Hall, 2007

J. Strong, *Concepts of Classical Optics*, New York: Dover, 2004 (Dover edition is an unabridged republication of the work originally published in 1958 by W. H. Freeman and Company, San Francisco, CA)

Specialized

Handbook of Optics, W. G. Driscoll (editor), W. Vaughan (associate editor), New York: McGraw-Hill, 1978

R. Ditteon, *Modern Geometrical Optics*, New York: John Wiley & Sons, 1998

D. C. O'Shea, *Elements of Modern Optical Design*, New York: Wiley, 1985

C. H. Palmer, *Optics: Experiments and Demonstrations*, Baltimore, MD: Johns Hopkins Press, 1962

W. J. Smith, *Modern Optical Engineering*, 3rd edition, New York: McGraw Hill, 2000

A. F. Wagner, *Experimental Optics*, New York: John Wiley & Sons, 1929

Chapters 2, 5, 6, 7, and 8 of this book

10 Simple Optical Instruments II: Light Illuminators and Microscopes

Objectives

1. Develop basic skills to make and manipulate simple optical instruments: Galilean telescope, light illuminators, and compound microscopes.
2. Determine the locations of the aperture stops, field stops, entrance and exit pupils, and entrance and exit windows of the studied systems.
3. Formulate general practical rules about how the performance of an optical instrument influences its image-making properties.

Background

Galilean Telescope

A Galilean telescope is formed by two lenses, one of which is a converging objective lens and the second is a diverging eye-lens. The focal length of the objective lens $f^{(1)}$ is much greater than the focal length $f^{(2)}$ of the eye-lens: $f^{(1)} \gg f^{(2)}$. Additionally, the back focal point of the object lens coincides with the back focal point of the eye-lens. The ratio $f^{(1)}/f^{(2)}$ determines the magnification of the telescope: $M = f^{(1)}/f^{(2)}$.

A Galilean telescope forms an erect image at infinity. A comparison of the two most widely used telescopes (refracting and Galilean) is shown in the following table.

Table 10.1

Telescope	Objective lens	Eye-lens	Distance between lenses[*]	Observed image	Magnification
Refracting telescope	converging	converging	$f^{(1)} + f^{(2)}$	Inverted	$f^{(1)}/f^{(2)}$
Galilean telescope	converging	diverging	$f^{(1)} - f^{(2)}$	Erect	$f^{(1)}/f^{(2)}$

[*] These expressions are correct for telescopes composed of simple thin lenses when front and back principal planes of the lenses coincide.

Light Illuminators

As a rule, the image produced by optical systems does not come from a self-luminous source. Objects are illuminated by light coming from light sources such as light bulbs. Of course, we can use some optics as a lens system (called a condenser) to collect light from the light source and send it in the exact direction of an object. After that, the so-called projecting optical system (lens system, mirrors, prisms, etc.) forms an image of the object. This image can be viewed on a screen or at infinity. In the following consideration, for simplicity, we will use the terms "condenser lens" to denote the optical system gathering light from the light source and "projection lens" to denote the lens system forming the image of the illuminated object on the screen. To illuminate an object, it should be placed between the condenser lens (or light source if not using a condenser) and the projection lens. The position of the object relative to both the light source and the projection lens is crucial and determines the type of illumination and magnification of the image on the screen. The most widely used schemes of illumination are described below.

(a) *Point source illumination* is where the object is placed between a point light source and a projection lens, as shown in Fig. 10.1. As can be seen from Fig. 10.1a, images of relatively small objects (their sizes are smaller than the projection lens aperture) have approximately uniform brightness. If the solid angle subtended by the light source image at the object image is comparatively small, we consider the brightness of the object image as uniform. When the size of the object is increased, the brightness of the object image changes greatly, becoming non-uniform. The central part of the object image has greater brightness than the edges of the image (Fig. 10.1b).

(b) *Collimated illumination* is where the object is placed between a condenser lens and a projection lens; the distance between the

Figure 10.1 Point source illumination: (a) The image of a small object (when object size is smaller than lens aperture) has approximately uniform brightness; (b) the image of the large object has non-uniform brightness; the central part of the image is brighter than the edges.

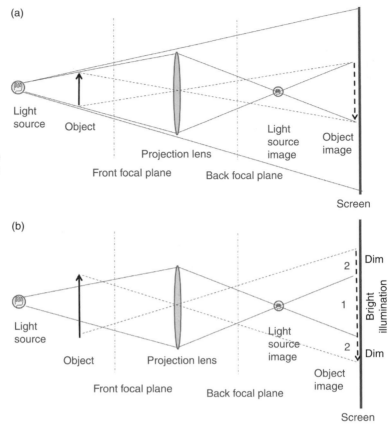

condenser lens and the point light source equals the front focal length $f^{(c)}_1$ of the condenser lens, as shown in Fig. 10.2. Collimated illumination improves uniformity of the image brightness. In addition, the size of an object for which the object images have uniform brightness can be increased (Fig. 10.2a). Collimated illumination still has regions of non-uniform image brightness (indicated by number 2 in Fig. 10.2b). To avoid regions of non-uniform brightness, the distance between the screen and the projection lens should be increased. However, in this case, total brightness of the image will be decreased.

(c) *Köhler illumination*, where the object is placed between a condenser lens and a projection lens with the position of the projection lens coinciding with the position of the light source image, is shown in Fig. 10.3. Köhler illumination produces object images of highly uniform brightness and is widely used in optical microscopy.

(a)

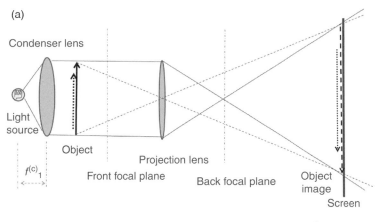

Figure 10.2
Collimated illumination: (a) Uniform brightness of images of both small and large objects; (b) non-uniformity of image brightness: 1 – region of uniform brightness, 2 – dim region.

(b)

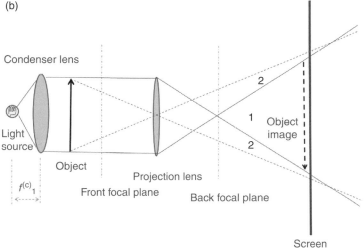

Compound Microscope

The compound microscope is used to magnify very fine objects that cannot be seen with the unaided eye or with a simple magnifier. The general optical scheme of the compound microscope is shown in Fig. 10.4. The microscope consists of an objective lens system (or, in other words, an objective) and eyepiece lens system (or ocular), which are separated by a distance L, as shown in Fig. 10.4. Objective lens systems (index "o") and eyepiece lens systems (index "e") are represented by their cardinal elements: Front (back) focal points $F^{(o,e)}_1$ ($F^{(o,e)}_2$); front (back) focal lengths $f^{(o,e)}_1$ ($f^{(o,e)}_2$); and front (back) principal points $H^{(o,e)}_1$ ($H^{(o,e)}_2$). Distance L is the distance between the back focal point of the objective lens system and the front focal point of the eyepiece lens system. Distance L is called the optical tube length and is assumed to

Figure 10.3 Optical scheme of Köhler illumination.

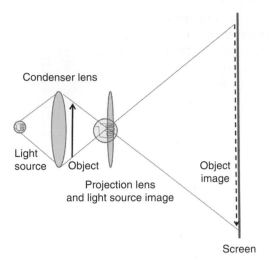

Figure 10.4 Principal optical scheme of the compound microscope.

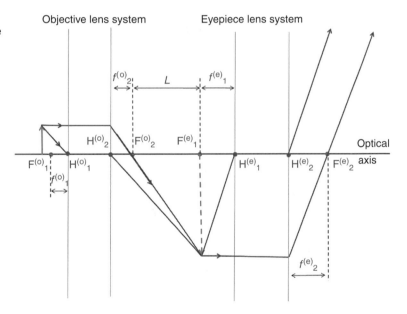

be equal to 160 mm. The object is placed in front of the objective lens system, at a distance between the focal and double-focal lengths of the microscope objective. The real intermediate image of the object, magnified by the objective, coincides with the front focal plane of the eyepiece lens system. The eyepiece lens system forms a magnified primary image at infinity, which can be observed with the human eye.

Total magnification M of the microscope can be found as the product of the magnification M_o of the microscope objective and the magnification M_e of the eyepiece lens system according to the following expressions (distances are measured in millimeters):

$$M_o = \frac{L}{f^{(o)}} \tag{10.1}$$

$$M_e = \frac{250}{f^{(e)}} \quad \text{image at infinity} \tag{10.2}$$

$$M_e = \frac{250}{f^{(e)}} + 1 \quad \text{image at normal near point} \tag{10.3}$$

$$M = M_o M_e = \begin{cases} \dfrac{L}{f^{(o)}} \dfrac{250}{f^{(e)}} \\ \dfrac{L}{f^{(o)}} \left(1 + \dfrac{250}{f^{(e)}} \right) \end{cases} \tag{10.4}$$

The numerical aperture (NA) is a measure of the light-gathering capability of the objective lens and can be defined by the formula $NA = n \sin \alpha$, where n is the refractive index of the medium immediately outside the objective, and α is half the angle subtended by the objective at the object. The numerical aperture also determines the resolving power of a microscope. This is its ability to form separate and distinct images of the fine elements of an object which are very close together. In the case of dry objects and objective, $n = 1$ and $NA = \sin \alpha$. The numerical aperture can be increased by immersing both the object and the objective in a drop of oil with the same refractive index as the optical material of the objective. In this chapter, we focus on dry objectives.

So far, we have considered the basic principles of the optical microscope, omitting practical realization of the optical scheme shown in Fig. 10.1. A real optical microscope consists of the following basic components: (1) A *light illuminator* (which can be represented by daylight and a mirror or a light bulb and a condenser as the gathering light lens system; (2) a *stage* to hold the specimen; (3) an *objective* (typical magnification values of objective lenses are $4\times$, $5\times$, $10\times$, $20\times$, $40\times$, $50\times$, $60\times$, and $100\times$); (4) an *ocular* (*eyepiece*) (typical magnification values for eyepieces include $2\times$, $5\times$, and $10\times$). All these components are assembled using a *frame*, or mechanical component of the microscope, holding all the optical elements and allowing adjustment of them. The performance of modern optical microscopes depends on their primary goal. For example, biological optical microscopes and polarizing optical microscopes differ greatly.

A *reticle* is an important component of an optical microscope, allowing size measurements of fine objects. A reticle is a net of fine lines

Materials Needed

- Optical table or optical rail
- Light sources (red cw laser, tungsten bulb)
- Object (arrow or cross)
- Ruler
- Mechanical holders
- Screws and screwdrivers
- Calipers and/or spherometer
- Squared graph paper
- Screen
- Needles or pins
- Mirror
- Thin lenses: Plano-concave, plano-convex
- Diaphragm (a simple opening made of paper may suffice) and adhesive tape to fix the diaphragm to a holder
- Reticle

or fibers placed at the focal plane of the eye-lens of the microscope. Depending on microscope performance, the reticle is placed at the front or at the back focal plane of the eye-lens. The word reticle comes from the Latin "reticulum," meaning "net." There are many variations of reticles; among them the most popular are crosshairs, precision scales, and angularly divided circular grids. Since the reticle is placed at the focal plane of the eye-lens, the reticle image is formed at infinity, coinciding with the primary image of the object. This allows us to use the reticle image as a ruler and to make size measurements of fine objects. Before such measurements, of course, the reticle should be calibrated.

Procedures

Wear gloves to handle optical elements (lenses).

Galilean Telescope

1. Identify two lenses in your optics kit according to the following description.

   Lens 1: Plano-convex, effective focal length $f^{(1)}$ ~100 mm, aperture diameter ~25 mm.

   Lens 2: Plano-concave, effective focal length $f^{(2)}$ ~20–25 mm, aperture diameter ~12 mm.

   Measure the focal lengths of the chosen lenses.

2. Assemble the experimental set-up shown in Fig. 10.5.

3. Align all the optical elements of the set-up as shown in Fig. 10.5a along the optical axis. Place the light source with an object (two crossed arrows) at distance $s \gg f^{(1)} + f^{(2)}$ as shown in Fig. 10.5a. Fix lens 1, and move lens 2 slowly; simultaneously observe the image of the object through lens 2 with your eye. Adjust the position of lens 2 to bring the image of the object in focus on the retina of your eye.

4. Note what type of image you see: erect or inverted?

5. Draw an optical scheme of the Galilean telescope using an appropriate scale, and show the aperture stop, entrance pupil,

(a)

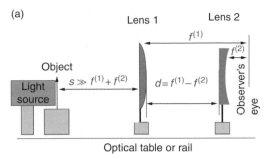

Lens 1   Lens 2

$f^{(1)}$

$f^{(2)}$

Object

$s \gg f^{(1)} + f^{(2)}$ $d = f^{(1)} - f^{(2)}$

Light source

Observer's eye

Optical table or rail

Figure 10.5 Galilean telescope.

(b)

exit pupil, field stop, entrance window, and exit window. Indicate all measured distances.

6. Place a diaphragm in front of the objective lens (lens 1). What changes do you notice in relation to the brightness of the image and the field of view? You may want to use a piece of adhesive tape to fix the diaphragm to the frame of the mounting holder placed in front of the lens. Do not place any tape directly onto a lens surface.

   Write your observations in a conclusions section.

7. Next, place this diaphragm in front of the eye-lens (lens 2). What changes do you notice in relation to the brightness of the image and the field of view? Write down your observations as a conclusion.

8. Calculate the magnification of the studied telescope and complete the following table.

Table 10.2

Distance between the lenses 1 and 2, d (mm)	Lens 1 focal length, $f^{(1)}$ (mm)	Lens 2 focal length, $f^{(2)}$ (mm)	Magnification, $M = f^{(1)}/f^{(2)}$

Point Source Illumination

Figure 10.6 Electrical circuit: Light source (incandescent lamp), resistor (10–15 Ω), and voltage supply (5 V) connected in series. The voltage supply should be able to handle an electrical current up to 1 A. Depending on the light source used, you may need to eliminate resistor R from this circuit to increase the amount of current going through the bulb and thus increase its brightness.

1. Identify a tungsten bulb available in your optics kit and design the electrical circuit shown in Fig. 10.6.

2. Assemble the set-up shown in Fig. 10.7. Place an object (arrow drawn on the piece of glass) between the screen and the tungsten bulb. Observe the image of the object on the screen and measure the distance s'_1 between the object and the image. Additionally, measure the height h_1 of the image. Increase the

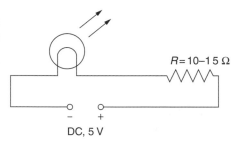

$R = 10–15 \ \Omega$

DC, 5 V

Figure 10.7 Point source illumination of the object (image is formed on the screen without a projection lens).

(a)

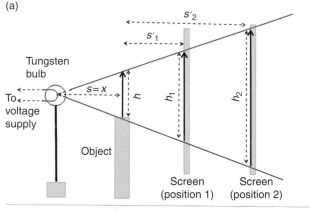

(b)

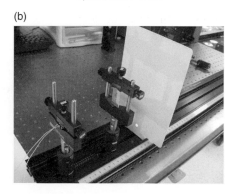

distance between the screen and the object and measure the image height h_2 and the new distance s'_2 between the object and the screen, as shown in Fig. 10.7.

3. Since the tungsten filament is located inside the bulb, you cannot measure the distance $s = x$ between the tungsten filament and the object. This distance can be found via measurements of the values of h_1, h_2, s'_1, and s'_2 using the following expression (derive it):

$$x = s = \frac{h_1 s'_2 - h_2 s'_1}{h_2 - h_1}$$

4. Complete the following table.

Table 10.3

Object height, h (mm)	Image height, h_1 (mm)	Distance, s'_1 (mm)	Image height, h_2 (mm)	Distance, s'_2 (mm)	Magnification, $M_1 = h_1/h$	Magnification, $M_2 = h_2/h$	Distance, $x = s$ (mm)

5. Increase the distance between the object and the screen and find the screen position for which the image is not in focus. For the found distance, qualitatively describe the brightness distribution of the image. Explain the obtained result.

6. Place a converging lens (focal length ~50 mm, lens aperture ~25 mm) between the object and the screen, as shown in Fig. 10.1. Bring the object image into focus on the screen by finding a new screen position. Additionally, find the image of the tungsten filament.

7. Observe the quality of the image brightness. Find the regions of largest and smallest brightness, and draw a conclusion about the spatial distribution of the image brightness.

Collimated Illumination

1. Identify a converging lens (effective focal length $f^{(c)}$ ~100 mm, aperture diameter ~25 mm) available in your optics kit.

2. Assemble the optical set-up shown in Fig. 10.8.

3. Place an object between the condenser lens and the screen. Measure the object height h, and the image heights h_1, h_2, and

Figure 10.8
Illumination of the object with collimated light.

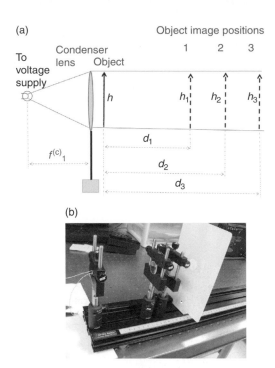

(a)

(b)

h_3 for different distances between the object and the screen d_1, d_2, and d_3, as shown in Fig. 10.8. It is recommended to keep d_1, d_2, and d_3 smaller than twice the focal length of the lens used. Complete the following table.

Table 10.4

Object height, h (mm)	Image height, h_1 (mm)	Distance, d_1 (mm)	Image height, h_2 (mm)	Distance, d_2 (mm)	Image height, h_3 (mm)	Distance, d_3 (mm)	Magnification, $M_1 = h_1/h$	Magnification, $M_2 = h_2/h$	Magnification, $M_3 = h_3/h$

4. Increase the distance between the object and the screen. Find the screen position for which the image is not in focus. Explain the obtained result.

5. Draw a conclusion about the brightness distribution of the image.

6. Place a converging lens (effective focal length $f^{(p)}$ ~30–50 mm, lens aperture ~35 mm) between the object and the screen as

(a)

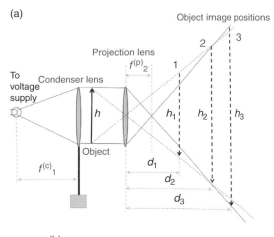

Figure 10.9 Optical scheme of collimated illumination.

(b)

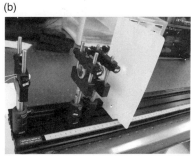

shown in Fig. 10.9. Find the image of the tungsten filament by moving the screen along the optical axis. Measure the distance between the found image of the lamp filament and the projection lens, and compare the measured value with the focal length of the projection lens. They should be the same.

7. Continue to move the screen along the optical axis until a magnified *inverted* image of the object on the screen is observed. Measure the image heights h_1, h_2, and h_3 for different distances between the projection lens and the screen d_1, d_2, and d_3, as shown in Fig. 10.9. For each position of the screen, calculate the magnification M, and complete the following table.

Table 10.5

Object height, h (mm)	Image height, h_1 (mm)	Distance, d_1 (mm)	Image height, h_2 (mm)	Distance, d_2 (mm)	Image height, h_3 (mm)	Distance, d_3 (mm)	Magnification, $M_1 = h_1/h$	Magnification, $M_2 = h_2/h$	Magnification, $M_3 = h_3/h$

8. Draw a conclusion about the brightness distribution of the magnified image.

Köhler Illumination

1. Identify two lenses in your optics kit, according to the following description:

   Lens 1 (to be the condenser lens): Plano-convex, effective focal length $f^{(c)}$ ~100 mm, aperture diameter ~25 mm.

   Lens 2 (to be the projection lens): Plano-convex, effective focal length $f^{(p)}$ ~100 mm, aperture diameter ~25 mm.

   Measure the focal lengths of the found lenses.

2. Assemble the optical set-up shown in Fig. 10.10. The principle of Köhler illumination requires that the positions of the projection lens and the image of the light source formed by the condenser lens should coincide, as shown in Fig. 10.10.

3. Place an object between the condenser lens and the projection lens, as shown in Fig. 10.10. Place the screen behind the projection lens (to the right) and find the image of the object on the screen. Measure the object height h, and the image

Figure 10.10 Optical scheme of Köhler illumination.

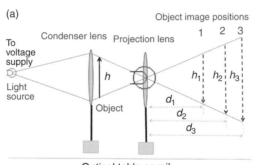

(a)

(b)

heights h_1, h_2, and h_3 for different distances between the projection lens and the screen d_1, d_2, and d_3, as shown in Fig. 10.10. Complete the following table.

Table 10.6

Object height, h (mm)	Image height, h_1 (mm)	Distance, d_1 (mm)	Image height, h_2 (mm)	Distance, d_2 (mm)	Image height, h_3 (mm)	Distance, d_3 (mm)	Magnification, $M_1 = h_1/h$	Magnification, $M_2 = h_2/h$	Magnification, $M_3 = h_3/h$

4. Increase the distance between the object and the screen and find the screen position for which the image is not in focus. Explain the obtained result.

5. Draw a conclusion about the brightness distribution of the image.

Compound Microscope (Collimated Illumination)

1. Identify three lenses in your optics kit according to the following description:

   Lens 1 (to be the condenser lens): Plano-convex, effective focal length $f^{(c)}$ ~100 mm, aperture diameter ~25 mm.

   Lens 2 (to be the objective lens): Plano-convex, effective focal length $f^{(o)}$ ~20–25 mm, aperture diameter ~35 mm.

   Lens 3 (to be the eyepiece lens): Plano-convex, effective focal length $f^{(e)}$ ~100 mm, aperture diameter ~25 mm.

   Measure the focal lengths of the lenses using any appropriate technique you know (refer to Chapter 6).

2. Assemble the optical set-up shown in Fig. 10.11.

3. Use a light box or an incandescent bulb as a light source. If you have no ground glass in your optics kit, you can make the equivalent of the ground glass by yourself. For example, adhesive tape attached to regular glass can be used instead of ground glass. For our purposes, the use of tape instead of ground glass is appropriate; however, such a replacement does not work in general (think why).

4. The distance between the ground glass and the condenser lens should be equal to the front focal length $f^{(c)}_1$ of the condenser lens. Make sure that the light output from the condenser lens is collimated.

Figure 10.11 Optical set-up of the compound microscope (collimated illumination). Remember that $L \sim 160$ mm.

(a)

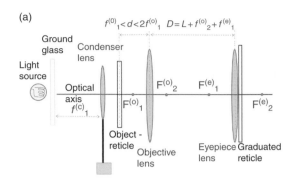

(b)

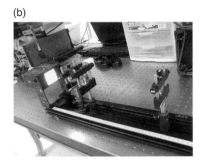

5. Place an object (square grid reticle) at a distance d from the objective lens: $f^{(o)}_1 < d < 2f^{(o)}_1$, where $f^{(o)}_1$ is the front focal length of the objective lens.

6. Place an eyepiece lens at a distance $D > f^{(o)}_2 + f^{(e)}_1$ behind the objective lens, as shown in Fig. 10.11; $f^{(o)}_2$ is the back focal length of the objective lens, and $f^{(e)}_1$ is the front focal length of the eyepiece lens.

7. Place a graduated reticle (X is the period of the reticle scale) in close contact with the eyepiece lens, as shown in Fig. 10.11.

8. Adjust the position of the objective lens until you observe the image of the object in focus at your eye.

9. Use a graduated reticle as a ruler and measure the period X_M of the magnified square grid reticle.

10. Measure the distance D between the objective lens and eyepiece lens.

11. Change the distance D and repeat the measurements (measure the period of the magnified square grid reticle). Measure the period X_0 of the non-magnified square grid reticle you used as an object. Figure 10.12 explains the meaning of the quantities X, X_M, X_0.

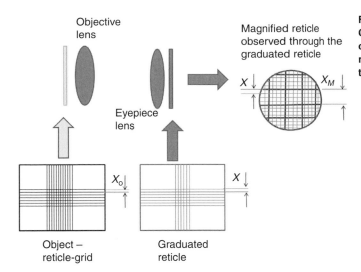

**Figure 10.12
Graduated reticle,
object reticle, and
magnified image of
the object reticle.**

12. Complete the following table.

Table 10.7

Focal length $f^{(o)}$ of objective lens (mm)	Focal length $f^{(e)}$ of eyepiece lens (mm)	Distance, D (mm)	Distance, L^* (mm)	Magnified reticle period, X_M (mm)	Non-magnified reticle period, X_o (mm)	Measured magnification, M^{**}	Calculated magnification, M^{***}
		100					
		150					
		200					
		235					
		250					

$^*\ L = D - f^{(o)} - f^{(e)}$
$^{**}\ M = X_M/X_o$
$^{***}\ M = M_o M_e = \frac{L}{f^{(o)}}\frac{250}{f^{(e)}}$

13. Plot the dependence of measured magnification M on distance L. Explain the obtained experimental results.

Compound Microscope (Köhler Illumination)

1. Identify three lenses in your optics kit according to the following description:
   Lens 1 (to be the condenser lens): Plano-convex, effective focal length $f^{(c)}$ ~100 mm, aperture diameter ~25 mm.

Figure 10.13 Optical set-up of the compound microscope (Köhler illumination).

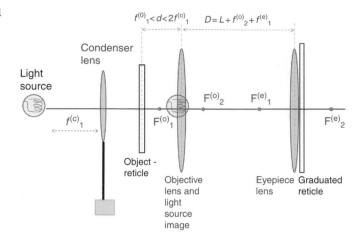

Lens 2 (to be the objective lens): Plano-convex, effective focal length $f^{(o)}$ ~20–25 mm, aperture diameter ~25 mm.

Lens 3 (to be the eyepiece lens): Plano-convex, effective focal length $f^{(e)}$ ~50 mm, aperture diameter ~25 mm.

Measure the focal length of these lenses using any appropriate technique you know.

2. Assemble the optical set-up shown in Fig. 10.13.

3. Use a light box or an incandescent bulb as the light source. Find the position of the magnified image of the light source and place the objective lens there, as shown in Fig. 10.13.

4. Place an object (square grid reticle) at a distance d from the objective lens: $f^{(o)}_1 < d < 2f^{(o)}_1$, where $f^{(o)}_1$ is the front focal length of the objective lens.

5. Place an eyepiece lens at a distance $D > f^{(o)}_2 + f^{(e)}_1$ behind the objective lens, as shown in Fig. 10.13. $f^{(o)}_2$ is the back focal length of the objective lens, and $f^{(e)}_1$ is the front focal length of the eyepiece lens.

6. Place a graduated reticle (X is the period of the reticle scale) in close contact with the eyepiece lens, as shown in Fig. 10.13.

7. Adjust the position of the objective lens until you observe the image of the object in focus with your eye. Observe changes in the object illumination. If the image is too bright for your eye, use a filter (attenuator) to decrease the output light power. Place the filter between the light source and the condenser lens.

8. Use a graduated reticle as a ruler and measure the period X_M of the magnified square grid reticle.

9. Measure the distance D between the objective lens and the eyepiece lens.

10. Change the distance D and repeat the measurements (measure the period of the magnified square grid reticle). Measure the period X_o of the non-magnified square grid reticle you used as an object. Figure 10.12 explains the meaning of the quantities X, X_M, and X_o.

11. Complete the following table.

Table 10.8

Focal length $f^{(o)}$ of objective lens (mm)	Focal length $f^{(e)}$ of eyepiece lens (mm)	Distance, D (mm)	Distance, L^* (mm)	Magnified reticle period, X_M (mm)	Non-magnified reticle period, X_o (mm)	Measured magnification, M^{**}	Calculated magnification, M^{***}
		100					
		150					
		200					
		235					
		250					

* $L = D - f^{(o)} - f^{(e)}$
** $M = X_M/X_o$
*** $M = M_o M_e = \frac{L}{f^{(o)}} \frac{250}{f^{(e)}}$

12. Plot the dependence of measured magnification, M on distance L. Explain the obtained experimental results.

13. Draw a conclusion about the improvements in the microscope performance due to using Köhler illumination.

Compound Microscope As a Measuring Device

1. Assemble the optical set-up shown in Fig. 10.14.

2. Place a reticle at the front focal plane of the eyepiece lens as shown in Fig. 10.14.

3. Place any graduated object (for example, you can measure the period of the grid reticle, then use it as a graduated object) between the condenser lens and the objective lens.

4. Adjust the position of the objective lens in order to bring the image of the graduated object into focus. In this case,

Figure 10.14 Optical microscope used as a measuring device.

(a)

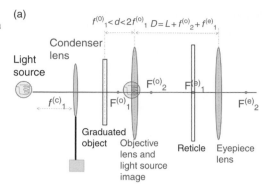

(b)

the image of the reticle and the image of the graduated object should coincide.

5. Measure the period of the reticle located at the front focal plane of the eyepiece using the image of the graduated object as a ruler (by looking at the eyepiece lens and observing images of the reticle and graduated object). Calculate the scale of the reticle. For example, if 10 divisions of the image of the graduated object correspond, let us say, to 5 divisions of the image of the reticle, we can conclude that 1 division of the reticle corresponds to the 2 divisions of the graduated object. The measurement period (the length of the one division) of the reticle is two times larger than the period of the graduated object. Therefore, by knowing the real value of the division length of the graduated object, we can calculate the period of the reticle.

6. Take any fine object and measure its size using the optical microscope.

Evaluation and Review Questions

1. Derive all the equations shown in the Background section.
2. Consider conditions where Köhler illumination is not feasible.

3. Choosing an appropriate scale, draw the aperture stops, field stops, pupils, and windows of the studied optical microscope.
4. Answer all the questions posed in the Procedure section.

For Further Investigation

Consider the Abbe scheme of illumination. The object is placed between a condenser lens and a projection lens in such a way that the object position coincides with the position of the light source image formed by the condenser lens. Draw an optical scheme of Abbe illumination and compare it with Köhler illumination.

Further Reading

General

E. Hecht, *Optics*, 4th edition, San Francisco, CA: Addison-Wesley, 2001

F. L. Pedrotti, S. J. L. Pedrotti, L. M. Pedrotti, *Introduction to Optics*, 3rd edition, Upper Saddle River, NJ: Pearson Prentice Hall, 2007

J. Strong, *Concepts of Classical Optics*, New York: Dover, 2004 (Dover edition is an unabridged republication of the work originally published in 1958 by W. H. Freeman and Company, San Francisco, CA)

Specialized

Handbook of Optics, W. G. Driscoll (editor), W. Vaughan (associate editor), New York: McGraw-Hill, 1978

R. Ditteon, *Modern Geometrical Optics*, New York: John Wiley & Sons, 1998

D. C. O'Shea, *Elements of Modern Optical Design*, New York: Wiley, 1985

C. H. Palmer, *Optics: Experiments and Demonstrations*, Baltimore, MD: Johns Hopkins Press, 1962

W. J. Smith, *Modern Optical Engineering*, 3rd edition, New York: McGraw Hill, 2000

Chapters 2, 5, 6, 7, and 8 of this book

11 Spherical Mirrors

Objectives

1. Develop basic skills to identify, classify, and handle spherical concave and convex mirrors.
2. Develop the basic skills needed to characterize spherical mirrors "at a glance."
3. Measure the focal lengths of concave and convex spherical mirrors using different experimental methods.
4. Determine the magnification produced by spherical mirrors.
5. Formulate the limitations of the experimental techniques studied.

Background

An ideal mirror can be defined as a surface which reflects 100 percent of the light falling on it. Mirrors can be classified depending on the shape of their reflecting surface. The most widely used mirrors are planar mirrors, spherical mirrors (concave and convex), and paraboloidal mirrors (concave and convex). Since, in the paraxial approximation, paraboloidal mirrors are governed by similar equations as spherical mirrors, let us describe spherical mirrors in more detail.

The reflecting surface of a spherical mirror has a spherical shape, which can be concave (Fig. 11.1a) or convex (Fig. 11.1b). When parallel light falls on a spherical mirror, it is reflected by the reflecting surface and converges at the focal point F as shown in Fig. 11.1.

The mirror's focal point F is located in the middle of the radius $R = OV$ of the spherical surface of the mirror: $OF = FV = OV/2$. The focal length of the mirror f is the distance between the focal point and the

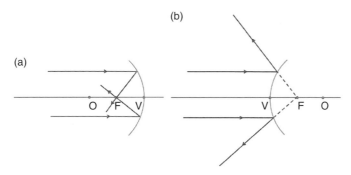

Figure 11.1 Spherical mirrors (V: vertex, F: focal point, O: center of curvature of the spherical surface): (a) Concave mirror, (b) convex mirror.

vertex V (the vertex can be determined as an intersection point of the spherical surface and the optical axis). Therefore, the absolute value of the mirror's focal length depends on the absolute value of the radius of curvature R of the mirror surface: $f = R/2$.

When an illuminated object is placed in front of the spherical mirror, the mirror produces an image of the object. The type of image (real or virtual, erect or inverted, magnified or minified) and the image distance s' (distance between the image and the mirror) depend strongly on the object distance s (distance between the object and the mirror) and the type of mirror used. The following table summarizes the imaging properties of spherical mirrors. All quantities in this table represent absolute values, i.e. they are *positive*.

Table 11.1

Concave mirror				
Object		**Image**		
Location	**Type**	**Location**	**Orientation**	**Relative size**
$s > 2f$	Real	$f < s' < 2f$	Inverted	Minified
$s = 2f$	Real	$s' = 2f$	Inverted	Same size
$f < s < 2f$	Real	$s' > 2f$	Inverted	Magnified
$s = f$		$\pm\infty$		
$s < f$	Virtual	$s' > s$	Erect	Magnified

Convex mirror				
Object		**Image**		
Location	**Type**	**Location**	**Orientation**	**Relative size**
Anywhere	Virtual	$s' < f$ $s > s'$	Erect	Minified

The object distance s, the image distance s', and the focal length f are governed by the mirror formula:

$$\frac{1}{s} + \frac{1}{s'} = \frac{1}{f} \tag{11.1}$$

It should be noted that this expression is written using the *sign convention* for spherical mirrors described briefly in the following table (we assume that the incident light travels from left to right).

Table 11.2

<table>
<tr><td colspan="4" align="center">Sign convention</td></tr>
<tr><td></td><td></td><td colspan="2" align="center">Sign</td></tr>
<tr><td>Quantity</td><td>Description</td><td align="center">+</td><td align="center">−</td></tr>
<tr><td>s</td><td>Object distance</td><td>Left of V, real object</td><td>Right of V, virtual object</td></tr>
<tr><td>s'</td><td>Image distance</td><td>Left of V, real image</td><td>Right of V, virtual image</td></tr>
<tr><td>f</td><td>Focal length</td><td>Concave mirror</td><td>Convex mirror</td></tr>
<tr><td>R</td><td>Radius of curvature</td><td>O right of V, convex</td><td>O left of V, concave</td></tr>
<tr><td>h</td><td>Object height</td><td>Above axis, erect object</td><td>Below axis, inverted object</td></tr>
<tr><td>h'</td><td>Image height</td><td>Above axis, erect image</td><td>Below axis, inverted image</td></tr>
</table>

Using the sign conventions, we can rewrite the mirror formula for the absolute (positive) values of the physical quantities (object distance $|s|$, the image distance $|s'|$, and the focal length $|f|$):

$$\frac{1}{|s|} + \frac{1}{|s'|} = \frac{1}{|f|} \quad \text{Concave mirror, object and image are real} \tag{11.2}$$

$$\frac{1}{|s|} - \frac{1}{|s'|} = -\frac{1}{|f|} \quad \text{Convex mirror} \tag{11.3}$$

Ray tracing (how to build an image) for spherical mirrors in the paraxial approximation is shown in Fig. 11.2.

Procedures

Wear gloves to handle optical elements (lenses).

"At-a-Glance" Spherical Mirror Characterization

(a) Qualitative Analysis

The type (concave or convex) of a spherical mirror can be determined by simple visual observation of the mirror shape. Additionally, by watching your own image produced by the spherical mirrors, you will notice that concave and convex mirrors act in different ways.

(a) Images produced by a convex mirror are all virtual, erect, and minified independently of the object position. This means that, by moving a convex mirror relative to any object, only the magnification (magnification M is the ratio of image height h' to object height h, $M = h'/h$) of the object (for example, your face) changes (of course, in all cases $M < 1$).

(b) Image parameters (real or virtual, erect or inverted, magnified or minified, value of magnification) of concave mirrors depend strongly on the relative positions of the mirror and object. By increasing the distance s between the object and the mirror,

Materials Needed

- Optical table or optical rail
- Light sources (red cw laser, tungsten bulb, LED)
- Object (arrow or cross)
- Ruler
- Mechanical holders; screws and screwdrivers
- Calipers and/or spherometer
- Squared graph paper
- Screen and screen with a hole; needles
- Concave and convex spherical mirrors
- Thin lenses: Plano-convex
- Diaphragm (a simple opening made of paper may suffice) and adhesive tape to fix the diaphragm to a holder

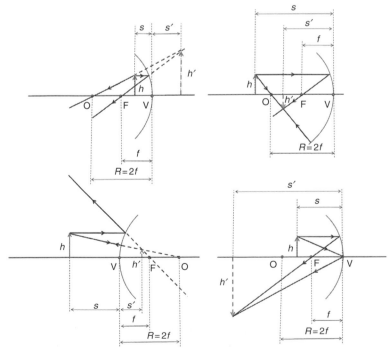

Figure 11.2 Ray tracing for concave and convex spherical mirrors.

the image will be virtual, erect, and magnified ($0 < s < f$); real, inverted, and magnified ($f < s < 2f$); real, inverted, and the same size ($s = 2f$); real, inverted, and minified ($s > 2f$).

(b) Quantitative Analysis

1. Analyze the spherical mirrors available in your optics kit. Determine the mirror surface shapes, then place a mirror in front of your eyes and watch the image of your face produced by the mirror. Notice the type of image and move the mirror slowly, increasing the distance between your eyes and the mirror. Simultaneously, observe changes in the image parameters. Draw a conclusion about the mirror type.

2. Place a reticle (a square or parallel grid printed on glass) in front of the mirror, as shown in Fig. 11.3. Since you do not know the focal length of the mirror, first place the reticle in close contact with the mirror, then increase the distance between the reticle and the mirror slowly. While the reticle is moving, observe the virtual images produced by the mirror. Compare the sizes of the image and the object. Locate the reticle position for which the image sizes are changed by a factor of 2: Twice magnified in the case of the concave mirror, and twice minified in the case of the convex mirror.

**Figure 11.3
Determination of
mirror focal length
using a reticle-grid:
Characterization of
(a) concave mirror
and (b) convex
mirror.**

3. Measure the distance s between the located reticle and the mirror. Calculate the mirror focal length according to the following expressions:

concave mirror: $M = 2$, $s = 0.5f \rightarrow f = 2s$

convex mirror: $M = 0.5$, $s = f \rightarrow f = s$

(a)

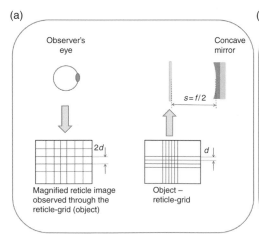

(b)

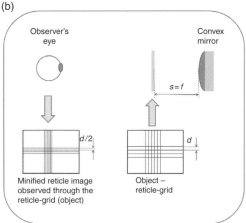

4. Measure the aperture diameter D of the mirrors, and complete the following table.

Table 11.3

Mirror	Magnification, M	Object distance, s (mm)	Focal length, f (mm)	Aperture diameter, D (mm)
Concave				
Convex				

5. Draw a conclusion about the limitations of the experimental technique used.

Determination of the Focal Length of a Concave Spherical Mirror Using Collimated Illumination

This method is based on a direct definition of the mirror focal point. Parallel light rays fall on a concave mirror and, after reflection from the mirror surface, all intersect in the focal point, or the position that is visually the brightest light spot.

1. Identify a concave mirror and a converging thin lens (focal length ~5–10 cm, aperture diameter ~25 mm) available in your optics kit.

2. Assemble the experimental set-up shown in Fig. 11.4.

3. Place the converging lens between the light source and the concave mirror in such a way that the distance between the light source and the converging lens equals the focal length of the lens. As a result, a parallel light beam will fall on the mirror.

4. Analyze the light reflected by the concave mirror and locate the position of the focal point F of the mirror. Hint: Reflected and

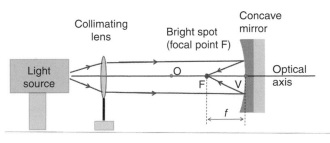

Figure 11.4 Determination of the focal point and the focal length of a concave mirror using collimated illumination.

139

falling light rays propagate in opposite directions; therefore, you cannot use a big screen to analyze reflected light. For these purposes, use a *small* piece of cardboard to prevent distortion and even blocking of the falling parallel light. Also, rotate the concave mirror slightly around the axis perpendicular to the optical axis of the optical system. This changes the direction of the reflected light beam so the path of the reflected light does not overlap the path of the falling light in the region of the focal plane.

5. Measure the distance between the detected focal point F and the mirror surface. The measured distance is the focal length of the mirror. Write down the measured value of the focal length:

   $f =$ _____ mm

6. Draw a conclusion about the limitations of the method used.

Determination of the Focal Length of a Convex Spherical Mirror Using Collimated Illumination

1. Identify a convex mirror and a converging thin lens (focal length ~5–10 cm, aperture diameter ~25 mm) available in your optics kit.

2. Assemble the experimental set-up shown in Fig. 11.5.

3. Place a converging lens between the light source and the convex mirror so that the distance between the light source and the converging lens equals the focal length of the lens. As a result, a parallel light beam will fall on the mirror, as shown in Fig. 11.5.

Figure 11.5
Determination of the focal point and the focal length of a convex mirror using collimated illumination.

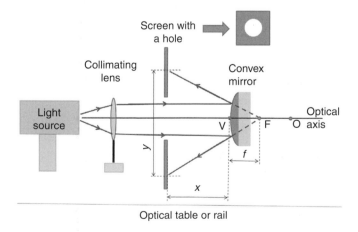

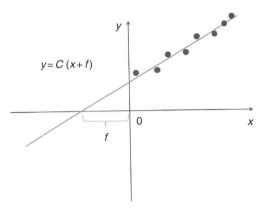

Figure 11.6
Determination of the focal length by fitting experimental data with a linear dependence.

$y = C(x + f)$

4. Place a screen with a hole to analyze the reflected light, as shown in Fig. 11.5. Measure distance x and width y of the reflected light beam (see Fig. 11.5 for explanations). Change the distance x and measure the corresponding value of the width y. Repeat such measurements for eight different positions x, and complete the following table.

Table 11.4

x (mm)								
y (mm)								

5. Plot the dependence of y on x. As can be shown directly from Fig. 11.6 (prove it), the expected dependence is $y = C(x + f)$ where C is a constant.

6. By fitting the plotted dependence with a straight line, determine the focal length of the convex mirror, as shown in Fig. 11.6.

7. Write down the measured value of the focal length:

   $f =$ _____ mm

8. Draw a conclusion about the limitations of the experimental technique you used.

Determination of the Focal Length of Concave and Convex Spherical Mirrors Using a Laser Beam

1. Use an expanded laser as a quasi-parallel light beam and repeat the measurements described in the two previous procedures.

Do this separately for the concave and the convex mirrors. Complete the following table.

Table 11.5

Concave mirror							
Focal length, f (mm)							
Convex mirror							
x (mm)							
y (mm)							
Focal length, f (mm)							

2. Compare the results obtained using an expanded laser beam and a collimated light beam. Draw a conclusion about the accuracy of the experiments for both cases and discuss how the size (width) of the parallel light beam used to measure the focal length of the spherical mirrors affects the experimental results.

Determination of the Focal Length of a Concave Spherical Mirror Using Point Source Illumination (Auto-collimation Method)

This method is based on the fact that a point light source placed at the center of curvature of a mirror and the image of the source formed by the mirror *coincide*. An obvious requirement to make good measurements is to use a bright and very small light source. In fact, the light sources used should be a *point* light sources. To make a point light source, use the *real and minified image* of a tungsten bulb (light source of finite size) produced by a thin lens.

1. Identify a concave mirror and converging thin lens (focal length ~5–10 cm, aperture diameter ~25 mm) available in your optics kit.

2. Assemble the experimental set-up shown in Fig. 11.7.

3. Place a thin converging lens at a distance $d > 2f_{lens}$ (f_{lens} is the focal length of the lens) from the light source, and locate the position of the minified light source image. Place a screen with a hole as shown in Fig. 11.7 (the position of the hole should coincide with the position of the light source image). In the following (for the current procedure), we refer to the light source image as a "point light source."

4. Place the concave mirror in close contact with the position of the point light source. Move the concave mirror slightly to increase the distance s between the point light source and the mirror.

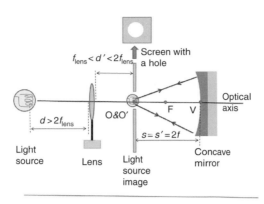

Simultaneously, watch the image of the point light source by using a small piece of cardboard as a testing screen. Hint: To observe the image of the point light source formed by the mirror, rotate the concave mirror slightly around the vertical axis perpendicular to the optical axis of the system. Locate the mirror position for which the point light source and its image coincide (or are of the same size). This position determines the center of curvature O of the mirror.

5. Measure the distance s between the center of curvature and the mirror surface. Calculate the focal length of the concave mirror $f = s/2$.

6. Write down the measured values:

   $s =$ _____ mm

   $f =$ _____ mm

7. Draw a conclusion about the limitations of the experimental technique you used.

Determination of the Focal Length of a Concave Spherical Mirror Using a Unit Magnification Method

1. Identify a concave mirror available in your optics kit.

2. Assemble the experimental set-up shown in Fig. 11.8.

3. Assuming you do not know the focal length of the mirror, make the distance between the light source and the concave mirror as long as you can. Place an object (arrow or cross) in the middle between the light source and the concave mirror, as shown in Fig. 11.8.

Figure 11.8 Unit magnification method to determine the focal length of a concave mirror.

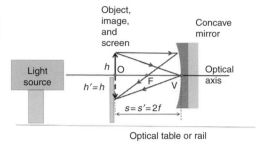

Optical table or rail

4. Use a small piece of cardboard to find the image of the object formed by the concave mirror. Ensure that the image is real, inverted, and magnified (otherwise you should increase the distance between the object and the mirror). Move the mirror towards the object (simultaneously observing the object image) until the image height equals the object height. Locate this position of unit magnification and measure the distance s between the object and the mirror.

5. Calculate the focal length of the concave mirror $f = s/2$.

6. Write down the measured values:

   $s =$ _____ mm

   $f =$ _____ mm

7. Draw a conclusion about the limitations of the experimental technique you used.

Determination of the Focal Length of a Concave Spherical Mirror and the Magnification Produced by the Same Mirror (Conjugate-Foci Method)

1. Identify a concave mirror available in your optics kit.

2. Assemble the experimental set-up shown in Fig. 11.9.

3. Make the distance between the light source and the concave mirror as long as you can. Place an object (arrow or cross) in close contact with the mirror, then move the object towards the light source. Use a small piece of cardboard to find the image of the object formed by the concave mirror. Locate the positions of the object and the object's image. Notice the type of image you observe – real, virtual, magnified, minified, erect, inverted.

4. Measure the distances s (object distance) and s' (image distance). In addition, measure the object height and image height. Calculate the magnification M produced by the mirror:

Figure 11.9

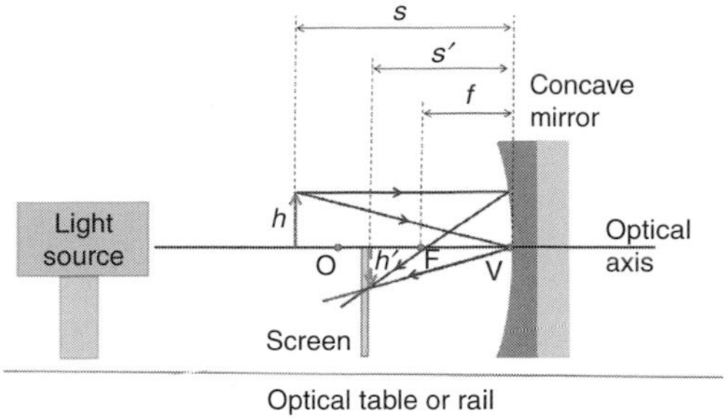

$M = h'/h$. Also, calculate the focal length f of the concave mirror by using the spherical mirror equation.

5. Write down the calculated values:

   $M =$ _____

   $f =$ _____ mm

6. Change the object distance (as shown in the table below), and complete the table.

Table 11.6

Object distance, s (mm)	Image distance, s' (mm)	Type of image	Object height, h (mm)	Image height, h' (mm)	Magnification M	Focal length, f^* (mm)
$>2f$						
$f<s<2f$						
$=2f$						

* Use Equations (11.1) and (11.2) to calculate the focal length of the mirror.

7. Draw a conclusion about the limitations of the experimental technique you used.

Determination of the Focal Length of a Convex Spherical Mirror Using an Auxiliary Convex Lens (Auto-collimation Method)

1. Identify a convex mirror and two converging thin lenses (focal length ~5 cm (lens 1) and ~10 cm (lens 2), aperture diameter ~25 mm) available in your optics kit.

2. Assemble the experimental set-up shown in Fig. 11.10.

3. Place a thin converging lens (#1, focal length ~5 cm) at a distance $d > 2f_{lens1}$ (f_{lens1} is focal length of lens 1) from the light source, and locate the position of the minified light source

Figure 11.10
Determination of the
focal length of a
convex mirror using
the auto-collimation
method.

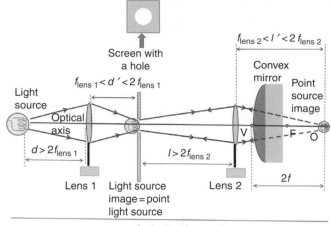

image. Place a screen with a hole as shown in Fig. 11.10 (the position of the hole should coincide with the position of the light source image). In the following stage, we will refer to the light source image as a "point light source."

4. Place a thin converging lens (#2, focal length ~10 cm) at a distance $l > 2f_{lens2}$ (f_{lens2} is focal length of lens 2) from the point light source, as shown in Fig. 11.10. Locate the position of the minified image O ("point source image" as shown in Fig. 11.10) of the point light source.

5. Place a convex mirror between lens 2 and the image of the point light source, as shown in Fig. 11. 10. The initial position of the convex mirror is very close to lens 2. Then, move the mirror towards position O, as shown in Fig. 11.10 (O is the image of the point light source formed by lens 2 in the absence of the convex mirror). Observe the light reflected by the mirror on the screen.

6. Move the mirror and locate the mirror position for which the light reflected by the mirror surface forms an auto-collimated image of the point light source. If the mirror position is located properly, the point light source coincides with the image of this source produced by the convex mirror.

7. Measure the distance VO between the convex mirror and the point source's image O, as shown in Fig. 11.10 (just remember – point source image O is the image of the point light source produced by lens 2 in the absence of the mirror).

8. Calculate the focal length of the convex mirror: $f = VO/2$.

9. Write down the measured values:

   VO = _____ mm

   f = _____ mm

10. Show that the experimental set-up shown in Fig. 11.10 makes sense only if the focal length of lens 2 is larger than the focal length f of the convex mirror: $f_{lens\ 2} > f$.

11. Draw a conclusion about the limitations of the experimental technique you used.

Evaluation and Review Questions

1. Derive all the equations shown in the Background section.
2. Compare the field of view of a convex mirror and a flat mirror, and explain why convex mirrors are used in vehicles.

Conclusion

Analyze your experimental results and complete the summarizing table.

Table 11.7

Methods used to determine focal length	Concave mirror	Convex mirror
"At-a-glance" method		
Collimated light beam		
Laser light beam		
Point light source		
Unit magnification method		N/A
Conjugate-foci method		N/A

For Further Investigation

Develop a search-light based on a concave spherical mirror. (Hint: Place the light source at the focal point of the mirror.)
Explain how the so-called telescopic mirror works.
Compare the performance of a metallic mirror and a dielectric mirror.

Further Reading

General

E. Hecht, *Optics*, 4th edition, San Francisco, CA: Addison-Wesley, 2001

F. L. Pedrotti, S. J. L. Pedrotti, L. M. Pedrotti, *Introduction to Optics*, 3rd edition, Upper Saddle River, NJ: Pearson Prentice Hall, 2007

J. Strong, *Concepts of Classical Optics*, New York: Dover, 2004 (Dover edition is an unabridged republication of the work originally published in 1958 by W. H. Freeman and Company, San Francisco, CA)

Specialized

Handbook of Optics, W. G. Driscoll (editor), W. Vaughan (associate editor), New York: McGraw-Hill, 1978

R. Ditteon, *Modern Geometrical Optics*, New York: John Wiley & Sons, 1998

D. C. O'Shea, *Elements of Modern Optical Design*, New York: Wiley, Wiley Series in Pure and Applied Optics, 1985

C. H. Palmer, *Optics: Experiments and Demonstrations*, Baltimore, MD: Johns Hopkins Press, 1962

W. J. Smith, *Modern Optical Engineering*, 3rd edition, New York: McGraw Hill, 2000

A. F. Wagner, *Experimental Optics*, New York: John Wiley & Sons, 1929

Chapters 2, 5, 6, 7, and 8 of this book

Introduction to Optical Aberrations

12

Objectives

1. Develop skills to make exact graphical ray tracings in order to build optical images.
2. Develop skills to classify and identify monochromatic optical aberrations: Spherical aberration, astigmatism, coma, Petzval field curvature, and distortion (qualitative analysis).
3. Develop skills to classify and identify chromatic aberrations (qualitative analysis).
4. Develop skills to measure and characterize optical aberrations (quantitative analysis).
5. Develop skills to minimize optical aberrations.
6. Formulate the limitations of the studied experimental techniques: Hartmann method, "squared paper" method, Foucault or "knife-edge" test.

Background

Exact Rays versus Paraxial Rays

We used the paraxial approximation (Gaussian, or first-order optics) to predict the imaging properties of optical systems. Paraxial optics uses a linear approximation of Snell's law $n_i \alpha_i = n_r \alpha_r$ instead of the exact expression $n_i \sin \alpha_i = n_r \sin \alpha_r$, where n_i is the refractive index of the incident medium, n_r is the refractive index of the refracting medium, α_i is the angle of incidence, α_r is the angle of refraction. Such an assumption is valid only for paraxial light rays, when the angles are arbitrarily small, and $\sin \alpha \approx \alpha$ (first-order approximation). Real optical systems are large enough that only a small part of all the light rays falls on the

Figure 12.1
Performance of a thick spherical lens for non-paraxial light rays: The position of the back focal point depends on the distance between the optical axis and the incident light ray (the case of spherical aberration).

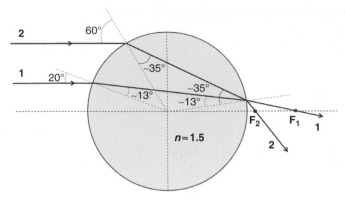

entrance pupil and so the rays can be considered paraxial. The remainder of the light rays (and this is the majority of light rays) are not paraxial: The real performance of an optical system deviates from a prediction based on the paraxial approximation. In the paraxial approximation, the rays which make an image do not necessarily meet at the paraxial image point, resulting in a blurred image, and the image being in the wrong position or of the wrong size. These faults, or defects, of images are called *optical aberrations*. Even the simple example shown in Fig. 12.1 demonstrates the obvious appearance of optical aberrations for non-paraxial light rays. Later you will see that this is the case of so-called spherical aberration. This thick spherical lens has different back focal points F_1 and F_2 for different light rays "1" and "2." Paraxial optics predicts only one back focal point for parallel rays falling on a lens.

Before we continue our discussion of optical aberrations, let us first introduce a few important definitions.

A *meridional ray* is a ray which is in a plane containing the optical axis (say, the z-axis of the chosen coordinate system). Any meridional ray remains in the same plane (the plane formed by the ray and the optical axis) as it passes through the optical system. The meridional ray begins in that specific plane and remains there as it passes through the system. Note that paraxial rays are meridional rays which are propagating very close to the optical axis.

Skew rays are any rays that are not meridional rays. This means that skew rays do not lie in a plane containing the optical axis. As a result, they will not remain in that plane while passing through the optical system, but will refract from plane to plane.

A *sagittal ray* is a skew ray which intersects the entrance pupil in a plane perpendicular to the meridional plane. We can also say that this is a ray which propagates in the plane which is perpendicular to the meridional plane and contains the chief ray.

Meridional rays and skew rays (which include sagittal rays) are called *exact rays*. This is to emphasize their *exact* origin (there is no approximation). The table shown below summarizes the classification of light rays.

Table 12.1

Light rays	
Paraxial rays	**Exact rays**
Rays which form *arbitrarily small* angles with an optical axis, therefore the *linear* approximation of Snell's law is valid: $n_i\alpha_i = n_r\alpha_r$	Rays which can form *any* angle with an optical axis and therefore the *general* form of Snell's law is required: $n_i \sin\alpha_i = n_r \sin\alpha_r$

Meridional rays	**Skew rays**
Rays which lie in the plane containing the optical axis	Any ray different from the meridional ray
Paraxial rays	**Sagittal rays**
Meridional rays which are propagating very close to the optical axis (and, as a result, the linear approximation of Snell's law can be used)	Skew rays which intersect the entrance pupil in a plane perpendicular to the meridional plane

Monochromatic Optical Aberrations

Paraxial optics can determine the image size and image location, but tells you nothing about the image quality. Image quality can be evaluated in a few ways. The most frequently used are *exact ray tracing* and the *third-order aberrational polynomial*.

(a) Exact ray tracing can predict *exact* image parameters (location, size, quality) via the direct application of Snell's law for each light ray falling on an optical surface. Since the time involved in tracing even one ray through a moderately complicated optical system is enormous, this method started to be commonly used only after the invention of the computer. Today, specially designed software (for example, Zemax) is widely used to design optical systems, and exact ray tracing has become the common method used by optical designers. We will consider elements of computer-based optical design in a separate chapter of this book.

(b) The difficulty in creating exact ray tracing (before computers were invented) stimulated scientists to develop an elegant theory of *approximations* of the exact calculations. This theory is based on the third-order approximation (*third-order optics* exploits the approximation $\sin\alpha \approx \alpha - \alpha^3/3!$), and results in two *aberration polynomials*. The derivation of the aberration polynomials is quite tedious and can be found in

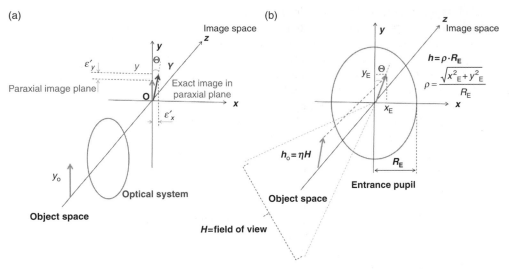

Figure 12.2 Paraxial and exact ray tracing for an optical system (R_E is the radius of the entrance pupil; H represents the full field of view, y_o and h_o are the object height; $\eta = y_o / H$ is the fractional object height; Θ is a polar angle measured from the y-axis; ρ is a coordinate defined as the ratio $\rho = \sqrt{x_E^2 + y_E^2}/R_E$, where x_E and y_E are the coordinates of the cross-section of the light ray and the entrance pupil).

specialized literature (see R. Ditteon's book for details). Therefore, we present only basic results of this theory.

An *aberration* is the difference in position between an exact ray and the corresponding paraxial ray. As shown in Fig. 12.2a, transverse components of the aberration can be written as follows:

$$\varepsilon_x' = X - x = X - 0 = X \tag{12.1}$$

$$\varepsilon_y' = Y - y \tag{12.2}$$

where x, y are the coordinates of the paraxial image and X, Y are the coordinates of the exact image in the paraxial plane.

The transverse components ε_x', ε_y' of the aberration are functions of ray position and direction, which can be expressed in polar coordinates, shown in Fig. 12.2b, as follows:

$$\varepsilon_x' = -\frac{1}{2n_k u_k}[S\rho^3 \sin\Theta + C\eta\rho^2 \sin 2\Theta + (A + P\Lambda^2)\eta^2\rho\sin\Theta] \tag{12.3}$$

$$\varepsilon_y' = -\frac{1}{2n_k u_k}[S\rho^3 \cos\Theta + C\eta\rho^2(2 + \cos 2\Theta) + (3A + P\Lambda^2)\eta^2\rho\cos\Theta + D\eta^3] \tag{12.4}$$

where η is the relative field position of the ray, ρ is the relative height of the ray in the entrance pupil, and Θ is the angle between the y-axis and the line connecting the center of the entrance pupil with the intersection point of the ray in the entrance pupil, n_k is the refractive index of image space, u_k is the slope angle of a marginal paraxial ray in image space, and

Λ is an optical invariant. An optical invariant is a measure of the light propagating through an optical system. It can be introduced in the following way. Consider two paraxial rays in a meridional plane passing through an optical system. The rays propagate through a medium characterized by the refractive index n. The angles between these two rays and an optical axis are $\bar{u}$ and u, respectively. These two rays cross an arbitrary plane which is perpendicular to the optical axis at two points. For each paraxial ray, the distance between the point of intersection and the optical axis is $\bar{h}$ and h, respectively. An optical invariant is defined as $\Lambda = n\bar{u}h - nu\bar{h}$. The optical invariant of two arbitrary meridional paraxial rays is a constant. By standard convention, the paraxial marginal and chief rays are used to calculate optical invariants.

The five coefficients S, C, A, P, and D are called the *Seidel aberration coefficients* and can be expressed as functions of the ray parameters for a full-field chief ray and a marginal axial ray. They depend on the system parameters and the paraxial ray parameters (slope angle of rays, object and image distances).

Each coefficient represents a certain type of optical aberration, as shown below:

Seidel aberration coefficient	Optical aberration
S	Spherical aberration
C	Coma
A	Astigmatism
P	Petzval curvature (curvature of the field)
D	Distortion

Let us discus each of the optical aberrations separately (but keep in mind that in a real experiment you can observe combinations of these optical aberrations. Each one is a player in the "aberration" team☺).

Spherical Aberration

Consider a set of parallel light rays falling on a converging lens. When light rays strike the surface at a distance greater than h above the optical axis, as shown in Fig. 12.3, they are focused at different focal points, which are located between paraxial focal point F_1 and the back vertex of the lens. By increasing the height h, the back focal point approaches the back vertex of the lens, as shown in Fig. 12.3. The distance between the axial intersection of a ray and the paraxial focal point is called the

Figure 12.3 Spherical aberration: LSA – longitudinal spherical aberration, TSA – transverse (lateral) spherical aberration.

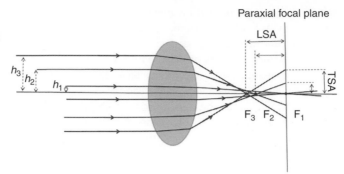

longitudinal spherical aberration (LSA). Convex lenses have positive LSA (as shown in Fig. 12.3), while diverging lenses have negative LSA. The height above the axis at which a given ray strikes the paraxial focal plane is called the transverse (or lateral) spherical aberration (TSA). The value of the spherical aberration is proportional to the square of the distance h between the optical axis and the light ray falling on the lens: LSA (or TSA) ~ h^2 (in the third-order optics approximation).

When a screen is placed in the paraxial focal plane, the bright paraxial focal point is surrounded by a symmetrical halo of dimmer light as a result of spherical aberration.

It should be noted that spherical aberration pertains only to object points that are on the optical axis (or are parallel to it).

Coma

Coma is an optical aberration which may take place when the object point is located at a large distance from the optical axis. In this case, the image of the point object is comet-like, as shown in Fig. 12.4.

Astigmatism

Pure coma is hard to display since it is usually combined with another off-axis aberration, astigmatism. When a point object is located at a finite distance from the optical axis, as shown in Fig. 12.5, meridional rays and sagittal rays will converge (after passing through the optical system) at two different points, F_T (converging point of the sagittal rays), and F_S (converging point of the meridional rays). As a result, two images (straight lines) will be formed. The primary image lies in the meridional plane, and the secondary image lies in the sagittal plane, as shown in Fig. 12.5. Between the primary and secondary images, the circle of least confusion is located. This is a circular blurred image of the point object. If we place a screen behind the secondary image, the image of the point object is an ellipse (see Fig. 12.5).

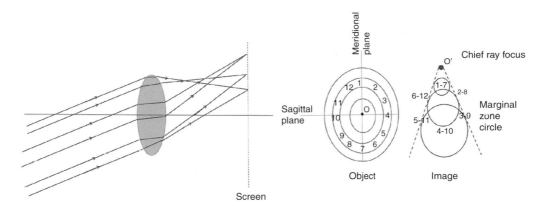

Figure 12.4 Coma.

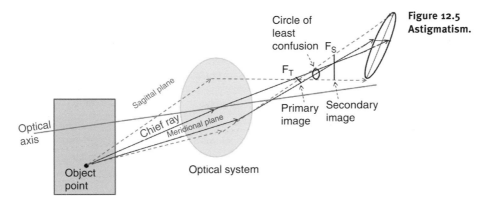

Figure 12.5 Astigmatism.

In summary, astigmatism deforms the image of a point object. Paraxial optics predicts a point image of a point object, but in reality the image of a point object is represented by an ellipse.

Petzval Field Curvature

Spherical aberration, coma, and astigmatism affect image quality and cause an image to blur. There are two additional types of optical aberrations. These are the Petzval field of curvature and distortion, which do not cause blurring, but influence the position and magnification of an image.

In image space, the flat object plane is represented by a curved (not flat!) image surface. This is the Petzval field curvature aberration. In other words, field curvature represents a longitudinal bending of the image field along the optical axis. This aberration can be seen even in a thin lens. It can be eliminated by using a combination of two lenses which satisfy the Petzval condition: $n_1f_1 + n_2f_2 = 0$. Optical systems which satisfy the Petzval condition have a flat field.

Figure 12.6
Distortion: (a) Object;
(b) image affected by
barrel distortion; (c)
image affected by pin-
cushion distortion.

(a)   (b)   (c)

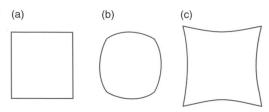

Distortion

Distortion occurs because the transverse magnification of an optical system can be a function of the off-axis image distance. There are two types of distortion: (1) Negative, or barrel distortion (magnification decreases with obliquity of the light rays); and (2) positive, or pincushion, distortion (magnification increases with obliquity of the light rays). Figure 12.6 demonstrates the effect of distortion on an image.

The following table summarizes data on optical aberrations and ways to eliminate them.

Table 12.2

Optical aberration	Location of object points	Feature/depends on	Ways to eliminate this aberration
Spherical aberration	On the optical axis	Can be positive or negative Shape of lens: ratio of radii of curvature of lens surfaces Lens orientation relative to falling light rays	(1) Combining the converging and a diverging lens – an achromatic doublet (2) Combining a lens and a mirror (3) Aplanatic points – a specific pair of conjugate points which are completely free of spherical aberration (4) Bending a lens: the process of changing the lens shape to reduce aberrations
Coma	Large distance from the optical axis	Can be positive or negative Shape of lens and angle of incidence	(1) Aperture stop at the proper location (2) Bending a lens – Coma can be made exactly zero for a single lens with a given object distance (~convex-planar lens)
Astigmatism	Large distance from the optical axis	Can be positive and negative; Shape of the lens and angle of incidence	Bending a lens in order to make the tangential and sagittal surfaces coincide – the resulting surface is called a Petzval surface
Field curvature	Large distance from the optical axis	Can be positive and negative	Petzval condition: $n_1 f_1 + n_2 f_2 = 0$ Field flattener lens Aperture stop
Distortion	Large distance from the optical axis	Can be positive and negative	Aperture stop

Procedures

Wear gloves to handle optical elements (lenses).

Longitudinal Spherical Aberration: Hartmann Method

(a) Converging Lenses

1. Identify converging lenses available in your optics kit according to the following description:

    Lens 1: Focal length ~12–18 mm, aperture diameter ~12 mm.

    Lens 2: Focal length ~100 mm, aperture diameter ~25 mm.

    Lens 3: Convex-convex, focal length ~10–30 mm, aperture diameter ~40–70 mm.

    Lens 4: Plano-convex, focal length ~50 mm, aperture diameter ~25 mm.

2. Assemble the laser beam expander according to the set-up shown in Fig. 12.7. To do this fix lens 1 and move lens 2 slowly until the back focal point of lens 1 and the front focal point of lens 2 coincide (Fig. 12.7). Observe the expanded laser beam on the screen. Make sure that your output laser beam is not converging or diverging. Use cardboard or a screen to check if the expanded laser beam is collimated.

Materials Needed
- Optical table or optical rail
- Light sources (red cw laser, tungsten bulb)
- Object (arrow or cross)
- Ruler
- Mechanical holders
- Screws and screwdrivers
- Calipers and/or spherometer
- Squared graph paper
- Screen and screen with a hole
- Needles (pins)
- Concave and convex spherical mirrors
- Thin and thick lenses: plano-convex, convex-convex, plano-concave
- Diaphragm (a simple opening made of paper may suffice) and adhesive tape to fix the diaphragm to a holder
- Hartmann screen
- Test reticle-grid
- Knife-edge

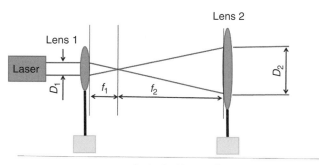

Figure 12.7 Optical set-up of a laser beam expander.

Optical table or rail

Figure 12.8 Optical set-up (a) to measure spherical aberration using a Hartmann screen (b).

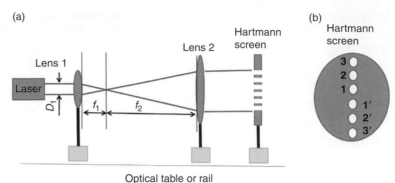

3. Place the Hartmann screen behind lens 2, as shown in Fig. 12.8. Use a cardboard screen and check to see if the light rays passing through the Hartmann screen are collimated.

4. Place the test lens (lens 3) behind the Hartmann screen, as shown in Fig. 12.9. Use a cardboard screen to locate the position of the back focal point for each pair of rays propagating at a distance h_i relative to the optical axis.

   Hint: To locate the back focal point F_i, use one pair of rays for each focal point. For example, if the Hartmann screen looks the same as the one shown in Fig. 12.8b, rays coming from the pair of holes 1 and 1' will converge to the back focal point F_1, and the rest of the back focal points can be found analogously. While determining a particular back focal point, block the remaining light rays by using a blocking screen.

5. Locate the paraxial back focal point of the lens and measure LSA (the distance between the real back focal point and the paraxial focal point) for each light ray propagating at a distance h_i from the optical axis, as shown in Fig. 12.9. Complete the following table.

   Table 12.3

h (mm)					
LSA (mm)					

6. Plot the dependence of LSA on h^2.

7. Draw a conclusion about the qualitative value of the spherical aberration and its dependence on falling ray parameters.

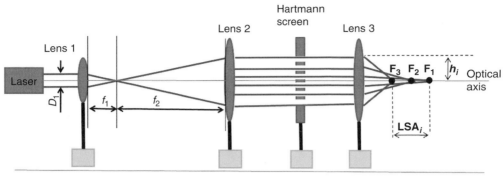

Optical table or rail

Figure 12.9
**Experimental set-up
of the Hartmann
method.**

(b) Spherical Aberration versus Lens Surface Shape

1. Replace lens 3 (convex-convex) with lens 4 (plano-convex).
   First place lens 4 in such a way that incident laser rays
   fall on the convex surface of the lens, as shown
   in Fig. 12.10.

2. Repeat all the steps you did with lens 3 with lens 4. Measure the
   dependence of LSA on h.

3. Rotate lens 4 about the vertical axis to orient it in such a way
   that incident laser rays fall on the flat surface of the lens, as
   shown in Fig. 12.11.

4. Repeat all the steps you made with lens 3 with lens 4. Measure
   the dependence of LSA on h. For both orientations of lens 4,
   complete the following table.

Table 12.4

Lens 4 "convex-planar"					
h (mm)					
LSA (mm)					
Lens 4 "planar-convex"					
h (mm)					
LSA (mm)					

5. Plot the dependence of LSA on h^2 for both orientations
   of lens 4.

6. Draw a conclusion about the dependence of spherical
   aberration on the shape.

Figure 12.10
Experimental set-up
to measure LSA of
lens 4.

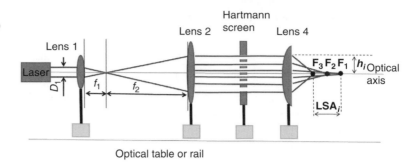

Optical table or rail

Figure 12.11
Experimental set-up
to measure LSA of
lens 4.

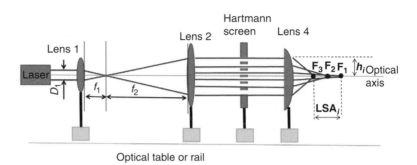

Optical table or rail

(c) Spherical Aberration of a Concave Spherical Mirror

Using the previous procedure, experimentally analyze the LSA of a concave mirror. Compare the obtained results for a mirror and a lens. Complete the following table and plot the dependence of LSA on h^2.

Table 12.5

h (mm)					
LSA (mm)					

Astigmatism

1. Identify converging lenses available in your optics kit according to the following description:

   Lens 1: Focal length ~ 2–18 mm, aperture diameter ~12 mm.

   Lens 2: Focal length ~100 mm, aperture diameter ~25 mm.

   Lens 3: Convex-convex, focal length ~10–30 mm, aperture diameter ~40–70 mm.

160

Lens 4: Plano-convex, focal length ~50 mm, aperture diameter ~25 mm.

2. Assemble the laser beam expander according to the set-up shown in Fig. 12.7.

3. Place the diaphragm so that you can observe the astigmatism behind lens 2, as shown in Fig. 12.12. Use a cardboard screen and see if the light rays passing through the diaphragm are collimated.

4. Place the test lens (lens 3) behind the diaphragm, as shown in Fig. 12.13.

5. First, make the angle between the falling rays and the optical axis of the lens equal to zero. Analyze the quality of the back focal point by means of the cardboard screen.

6. Rotate lens 3 about the vertical axis, as shown in Fig. 12.13. Notice the appearance of astigmatism (use the cardboard screen to locate the aberrated focal point).

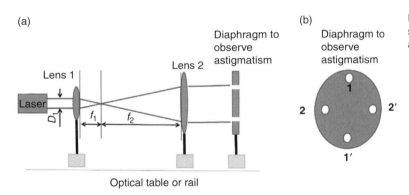

Figure 12.12 Optical set-up to demonstrate astigmatism.

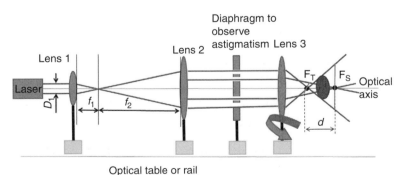

Figure 12.13 Optical set-up to demonstrate astigmatism.

7. Locate the primary image, the circle of least confusion, and the secondary image. If possible, measure the distance $F_T F_S$, as shown in Fig. 12.13 for different orientations of lens 3 for angles between the falling rays and the normal of the lens surface. Notice changes in image quality while rotating lens 3.

8. Exchange lens 3 for lens 4 and repeat the same procedures as completed with lens 3. First, place lens 4 in such a way that the incident laser rays fall on the convex surface of the lens (as was shown in Fig. 12.10). Locate the primary image, the circle of least confusion, and the secondary image. If possible, measure the distance $F_T F_S$, as shown in Fig. 12.13, for the different orientations of lens 4 (for the angles between the falling rays and the normal of the lens surface). Notice changes in the image quality while rotating lens 4.

9. Rotate lens 4 by 180° and repeat all the measurements and observations for this position of the lens.

10. For lenses 3 and 4, complete the following table.

Table 12.6

	Lens 4 "convex-planar"	Lens 4 "planar-convex"	Lens 3
Angle of incidence			
$F_T F_S$ (mm)			

11. Do you observe pure astigmatism or a combination of the optical aberrations?

Curvature of the Field and Distortion

Place the large reticle-grid between a point light source and a lens (choose any lens available in your optics kit).

By moving the screen along the optical axis, find the image of the reticle-grid on the screen.

Analyze the quality of the magnification and determine the type of distortion (barrel or pin-cushion).

Compare the image of the central part of the reticle and the image of the edges. If the image of the reticle edge is not in focus, move the screen slightly to bring the image of the reticle edge into focus. Draw a conclusion about the curvature of the field.

Change the lens and repeat the same procedures. Complete the following table.

Table 12.7

Lens	Type of distortion

Evaluation and Review Questions

1. Develop an experimental procedure to measure the spherical aberration of a diverging lens.
2. Develop an experimental procedure to observe chromatic aberration experimentally.

Conclusion

Analyze the obtained experimental results and draw a conclusion about the contribution of optical aberrations to the final quality of an image.

For Further Investigation

Find and read about interference methods to determine optical aberrations.

Describe the Foucault or "knife-edge" test.

Further Reading

General

E. Hecht, *Optics*, 4th edition, San Francisco, CA: Addison-Wesley, 2001

F. L. Pedrotti, S. J. L. Pedrotti, L. M. Pedrotti, *Introduction to Optics*, 3rd edition, Upper Saddle River, NJ: Pearson Prentice Hall, 2007

J. Strong, *Concepts of Classical Optics*, New York: Dover, 2004 (Dover edition is an unabridged republication of the work originally published in 1958 by W. H. Freeman and Company, San Francisco, CA)

Specialized

Handbook of Optics, W. G. Driscoll (editor), W. Vaughan (associate editor), New York: McGraw-Hill, 1978

R. Ditteon, *Modern Geometrical Optics*, New York: John Wiley & Sons, 1998

D. C. O'Shea, *Elements of Modern Optical Design*, New York: Wiley, 1985

C. H. Palmer, *Optics: Experiments and Demonstrations*, Baltimore: Johns Hopkins Press, 1962

W. J. Smith, *Modern Optical Engineering*, 3rd edition, New York: McGraw Hill, 2000

A. F. Wagner, *Experimental Optics*, New York: John Wiley & Sons, 1929

Chapters 2, 5, 6, 7, 8, and 11 of this book

Elements of Optical Radiometry

13

Objectives

1. Measure the total power of the light emitted by an incandescent lamp.
2. Measure the irradiance of a screen due to light produced by a real light source.
3. Observe vignetting.
4. Make and characterize a quasi-Lambertian light source.

Background

Radiometry is a subfield of optical science dealing with methods to measure optical radiant energy. In short, radiometry deals with measurements of radiometric quantities such as radiant energy and radiant flux. The most widely used radiometric quantities and their definition are shown in Table 2A.1, fully reproduced in this chapter.

In many practical cases, the sensitivity of the human eye should be taken into account. This can be done by considering the area of photometry. The relationships between basic radiometric and photometric units are shown in Table 2A.1.

This chapter focuses on measurements of radiometric quantities using the simple optical detector described in Chapter 4. After completing the experimental tasks in this chapter, you will appreciate the art and science of optical radiometry and its importance to numerous applications, including the booming field of light-illuminating systems.

Table 2A.1 Basic radiometric and photometric characteristics

| Basic concept | Spatial density | | | |
	Areal density at a surface	Intensity	Specific intensity	Volumetric density
Radiometric units				
Radiant energy, joules, J				Radiant density, joules per cubic meter, J/m^3
Radiant flux, watts, W	Radiant exitance, irradiance, watts per square meter, W/m^2	Radiant intensity, watts per steradian, W/sr	Radiance, watts per steradian-square meter, $W/(sr\ m^2)$	
Photometric units				
Luminous energy, lumen seconds [talbot], lm s				Luminous density, lumen seconds per cubic meter, $lm\ s/m^3$
Luminous flux, lumens, lm	Luminous exitance, illuminance, lux (lx), lumens per square meter, lm/m^2	Luminous intensity, candela, cd [lumens per steradian], lm/sr	Luminance [photometric brightness], $lm/(sr\ m^2)=cd/m^2$	
Definitions				
Power emitted by a light source into the whole space per time unit	Power per detector area	Power emitted by a light source into a solid angle	Power emitted by a Lambert irradiator of a surface into a solid angle	Energy per unit volume

Unit Conversions

Radiant flux: 1 W (watt) = 683.0 lm at 555 nm
              1 J (joule) = 1 W s (watt second)

Luminous flux: 1 lm (lumen) = 1.464×10^{-3} W at 555 nm = $1/(4\pi)$ cd (candela) (only if isotropic)
              1 lm s (lumen second) = 1 talbot (T) = 1.464×10^{-3} J at 555 nm

Procedures

Wear gloves to handle optical elements (lenses).

    If you have not already done so, design an electrical circuit as shown in Fig. 13.1 to make a linear light detector (see Chapter 4 for more details).

How to Measure Total Light Power Emitted by a Light Source

1. First, make sure that the light detector you are using detects light power, not irradiance. To do this, assemble the simple optical scheme shown in Fig. 13.2. (a) Use a laser to irradiate the light detector and measure the detector's response (V_{out} as shown in Fig. 13.1). If the light signal saturates the detector, use attenuators to decrease the power of the light. (b) Insert a lens between the light source (laser) and the light detector in order to increase the laser beam diameter, as shown in Fig. 13.2b. Measure the detector response again. Ensure that all the light is collected by the detector. To do this, find the right position for the lens by moving the lens.

2. Estimate the diameter D of the light beam falling on the detector for both cases (a) and (b), and complete the following table.

Table 13.1

Light power, P (a.u.[*])	Light power, P (mW)[**]	Light beam diameter, D (mm)	Irradiance, E_e[***] (a.u.)	Irradiance, E_e[***] (mW/mm²)	Detected signal, V_{out} (V)
1					
1					

[*] a.u. – arbitrary units
[**] specified at the time of the experiment
[***] $E_e = \frac{P}{\pi(D/2)^2}$

3. Draw a conclusion about "what physical quantity the light detector detects."

4. Calibrate your light detector.

5. Identify an incandescent lamp in your optics kit and design the electrical circuit shown in Fig. 13.3. Simply connect the incandescent lamp to the DC outlet "5 V 5 A."

Figure 13.1 Electrical circuit of light detector based on a photodiode: Photoconductive (reverse-biased) mode of operation with operational amplifier.

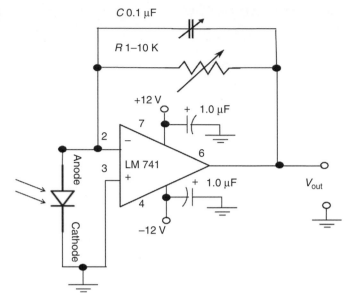

Figure 13.2 Optical scheme to verify what physical quantity a light detector detects.

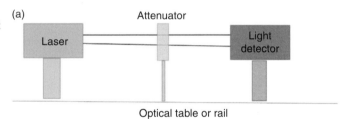

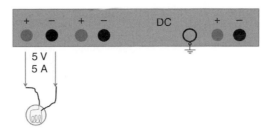

Figure 13.3
Incandescent light
source.

5 V
5 A

(a)

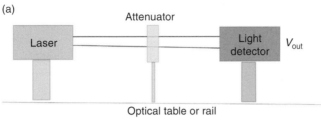

Figure 13.4 Optical
scheme to measure
total light power
emitted by an
incandescent bulb.

Attenuator

Laser

Light
detector

V_{out}

Optical table or rail

(b)

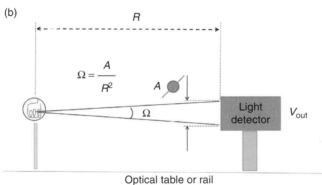

R

$$\Omega = \frac{A}{R^2}$$

A

Ω

Light
detector

V_{out}

Optical table or rail

6. Assemble the optical set-up shown in Fig. 13.4a. Choose the right attenuator by trial to reach the detector response $V_{out} \sim 8-10$ V. Write down the measured value of V_{out}_____V.

7. Assemble the optical set-up shown in Fig. 13.4b. Adjust the distance R between the light source and the detector to reach the same detector response V_{out} as you measured in step 6.

8. Measure the distance R between the light source and the light detector, and the area A of the photosensitive surface of the light detector.

9. Complete the following table.

Table 13.2

R (mm)	A (mm²)	Solid angle, $\Omega = A/R^2$	V_{out} (V)	Attenuation, T (%)	Detected laser power, P^* (mW)	Total light power emitted by bulb, P_t^{**} (mW)

* $P = TP_{max}/100$, where P_{max} is maximal laser output light power (specified at the time of the experiment)
** $P_t = (P/\Omega)4\pi$

10. Draw a conclusion about the limitations of the technique described above.

Irradiance of a Screen by a Real Light Source

1. Assemble an optical set-up as shown in Fig. 13.5.

2. Use a stack of two or three rods to hold the incandescent bulb and detector to provide large values of the plane angle Θ (30–60°). Fix the distance D between the light source and the screen and move the light detector along the horizontal axis. This changes the distance h as shown in Fig. 13.5. For each position of the detector, measure the distance h and measure the light signal V_{out}.

3. Change the distance D, and repeat the measurements described in step 2. Complete the following table.

Table 13.3

Distance D (mm)	Distance h (mm)	cos Θ	cos³Θ	cos⁴Θ	V_{out} (V)

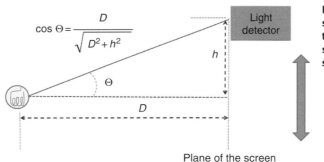

Figure 13.5 Optical scheme to measure the irradiance of a screen by a real light source.

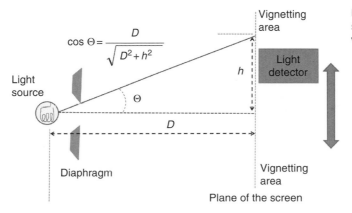

Figure 13.6 Optical scheme to observe vignetting.

4. For each distance D, plot V_{out} vs. $\cos^3\Theta$ and V_{out} vs. $\cos^4\Theta$. Which of the dependences better describes the experimental results?

5. Explain the obtained experimental results.

Vignetting

1. Assemble an optical set-up as shown in Fig. 13.6.

2. Adjust the distance D between the light source and the screen to observe vignetting – a rapid darkening of the edges of the light spot on the screen formed by light passing through the diaphragm. Make sure that the size of the vignetting area is large enough to measure its irradiance. For your distance D, measure the irradiance of the screen at different positions (different distances h, as shown in Fig. 13.6). Measure the irradiance for at least two positions of the light detector located in the vignetting area.

3. Change the distance D and repeat the measurements described in step 2. Complete the following table.

Table 13.4

Distance D (mm)	Distance h (mm)	$\cos \Theta$	$\cos^3 \Theta$	$\cos^4 \Theta$	V_{out} (V)

4. For each distance D, plot V_{out} vs. $\cos^3 \Theta$ and V_{out} vs. $\cos^4 \Theta$. Which of the dependences better describes the experimental results? Find the angles of the vignetting from your plots.

5. Explain the obtained experimental results.

How to Make a Quasi-Lambertian Light Source

1. Assemble the optical set-up shown in Fig. 13.7.

2. Make sure that all distances R are equal to each other. For each angle Θ, measure the irradiance by using the light detector as shown in Fig. 13.7 (V_{out}).

3. Complete the following table.

Table 13.5

Distance R (mm)	Angle Θ	$\cos \Theta$	V_{out} (V)

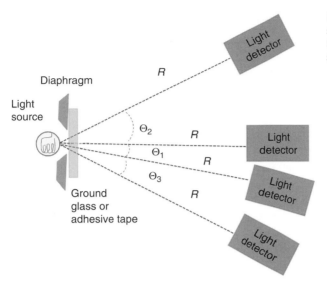

Figure 13.7 Optical scheme of the quasi-Lambertian light source.

4. Plot V_{out} vs. $\cos\Theta$ and draw a conclusion about the light source you studied.

Conclusion

Analyze the obtained experimental results.

For Further Investigation

Read more about point and extended light sources, the concept of etendue, the brightness theorem, luminous efficiency, and devices used to perform photometric measurements (they are called photometers, for example, the Lummer–Brodhun type of photometer).

Further Reading

General

E. Hecht, *Optics*, 4th edition, San Francisco, CA: Addison-Wesley, 2001

F. L. Pedrotti, S. J. L. Pedrotti, L. M. Pedrotti, *Introduction to Optics*, 3rd edition, Upper Saddle River, NJ: Pearson Prentice Hall, 2007

J. Strong, *Concepts of Classical Optics*, New York: Dover, 2004 (Dover edition is an unabridged republication of the work originally published in 1958 by W. H. Freeman and Company, San Francisco, CA)

Specialized

Handbook of Optics, W. G. Driscoll (editor), W. Vaughan (associate editor), New York: McGraw-Hill, 1978

R. Ditteon, *Modern Geometrical Optics*, New York: John Wiley & Sons, 1998

D. C. O'Shea, *Elements of Modern Optical Design*, New York: Wiley, 1985

C. H. Palmer, *Optics: Experiments and Demonstrations*, Baltimore, MD: Johns Hopkins Press, 1962

W. J. Smith, *Modern Optical Engineering*, 3rd edition, New York: McGraw Hill, 2000

A. F. Wagner, *Experimental Optics*, New York: John Wiley & Sons, 1929

Chapters 2, 5, 6, 7, 8, 9, and 11 of this book

Cylindrical Lenses and Vials

14

Background

Cylindrical Lenses

Cylindrical lenses (Fig. 14.1) are special types of lenses used in the field of optometry to correct astigmatism, in novel visual displays, and in many other modern optical applications. A cylindrical lens lacks rotational symmetry about its optical axis and, as a consequence, has asymmetric focusing properties. As we already know, the ideal spherical lens produces a point image of a point object (for this reason a spherical lens is said to be *stigmatic*). In contrast to this, a cylindrical lens produces a line image of a point object. Because of this, a cylindrical lens is *astigmatic*.

Consider the cross-sections of a cylindrical lens in planes XZ and YZ as shown in Fig. 14.1. The cross-section in plane YZ is rectangular, while the cross-section in plane XZ is part of a circle. These cross-sections

**Figure 14.1
Cylindrical lens.**

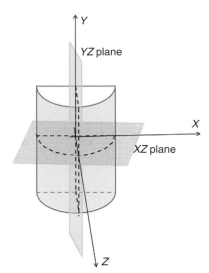

**Figure 14.2 Cross-
sections of a
cylindrical lens and
their imaging
properties: (a) *XZ*
cross-section acts as
a spherical lens;
(b) *YZ* cross-section
acts as a plane
parallel plate.**

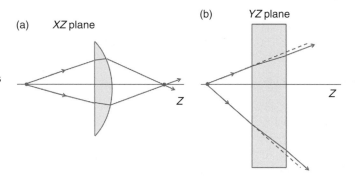

have different imaging properties. If we consider a point light source in the *XZ* and *YZ* planes, then the *XZ* cross-section acts as a spherical lens (Fig. 14.2a) and the *YZ* cross-section as a plane parallel plate (Fig. 14.2b).

Now we understand why a cylindrical lens produces a *line image* of a *point light source*, as shown in Fig. 14.3.

It should be noted that all the equations derived for spherical lenses can also be applied to cylindrical lenses, but *only for the plane cross-section* (*XZ* shown in Figs. 14.1–14.3). With a certain degree of caution, the information you applied to spherical lenses and parallel plates can also be applied in experiments with cylindrical lenses.

Cylindrical Vials

Cylindrical vials are widely used in our everyday lives. Most people use cylindrical vials several times a day to drink coffee, tea, or other liquids.

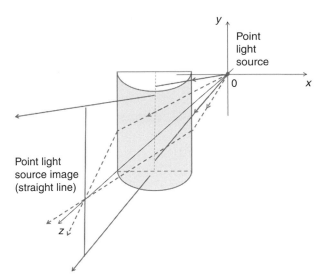

Figure 14.3 Imaging of a point light source by a cylindrical lens.

In addition to the obvious importance of cylindrical vials in our daily lives, they have interesting applications in geometrical optics.

(a) Empty Cylindrical Vial

Consider an empty cylindrical vial formed by two coaxial cylinders of different radii: Internal radius of the cylinder is R_1 and external radius is R_2. Suppose that the space between the two cylinders is filled homogeneously with optical material of refractive index n (Fig. 14.4a).

Now consider a light ray falling on the cylindrical vial from the left to the right, as shown in Fig. 14.4a. Exact ray tracing for this light ray is made and is also shown in Fig. 14.4a (numerical values of the angles $\alpha_0 \ldots \alpha_6$ are presented in Table 14.1, assuming that the refractive index of the vial material is 1.5). As can be seen, an *empty cylindrical vial, as shown in Fig. 14.4a, acts as a diverging cylindrical lens*.

Figure 14.4 allows us to do the following simple calculations:

$$OW = R_1; OV = OM = R_2; FV = f$$

$$\alpha = \pi - [\alpha_0 + 2(\alpha_2 - \alpha_1) + (\pi - 2\alpha_3)] = 2(\alpha_3 + \alpha_1 - \alpha_2) - \alpha_0 \quad (14.1)$$

$$\beta = \pi - [\alpha_0 + \pi - \alpha] = \alpha - \alpha_0 \quad (14.2)$$

$$\tan(\beta) = \frac{MN}{FN} = \frac{OM \sin \alpha}{FV - NV} = \frac{OM \sin \alpha}{FV - (OV - OM \cos \alpha)} = \frac{R_2 \sin \alpha}{f - R_2(1 - \cos \alpha)} \quad (14.3)$$

Table 14.1 Numerical values of angles $\alpha_0 \ldots \alpha_6$ shown in Fig. 14.3 (refractive index is assumed to be 1.5)

α_0	$32°$
α_1	$21°$
α_2	$31°$
α_3	$51°$
α_4	$51°$
α_5	$31°$
α_6	$21°$
α_7	$32°$

Figure 14.4 Empty cylindrical vial acts as a diverging cylindrical lens.

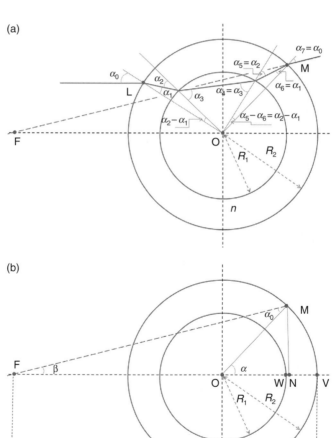

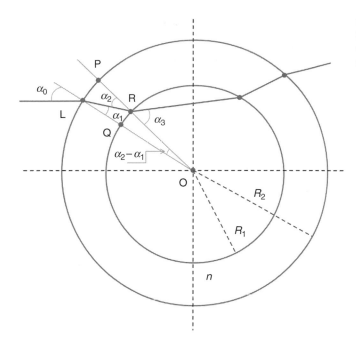

Figure 14.5
Schematic drawing
which helps to prove
expression (14.5).

$$\frac{\sin(\alpha_0)}{\sin(\alpha_1)} = \frac{\sin(\alpha_3)}{\sin(\alpha_2)} = n \qquad (14.4)$$

To simplify expression (14.3), let us consider the paraxial approximation, $\tan\beta \approx \beta$; $\sin\alpha \approx \alpha$; $\cos\alpha \approx 1$.

For the paraxial approximation, the following expression can be used:

$$\frac{\alpha_2}{\alpha_1} = \frac{R_2}{R_1} \qquad (14.5)$$

This expression can be proved using Fig. 14.5, as shown below:

$$\alpha_2 \approx \tan\alpha_2 = \frac{LP}{PR} = \frac{R_2(\alpha_2 - \alpha_1)}{R_2 - R_1} \qquad (14.6)$$

$$\alpha_1 \approx \tan\alpha_1 = \frac{QR}{LQ} = \frac{R_1(\alpha_2 - \alpha_1)}{R_2 - R_1} \qquad (14.7)$$

From expressions (14.6) and (14.7), expression (14.5) is derived by means of direct division.

By using the paraxial approximation and expressions (14.1), (14.2), (14.4), and (14.5), we can rewrite expression (14.3) in the following form (we present the entire derivation since this topic is hard to find in the literature):

$$\beta \approx \frac{R_2 \alpha}{f} \tag{14.8}$$

Substitute β by using expressions (14.1) and (14.2), and rewrite (14.8):

$$2(\alpha_3 + \alpha_1 - \alpha_2 - \alpha_0) = \frac{R_2[2(\alpha_3 + \alpha_1 - \alpha_2) - \alpha_0]}{f} \tag{14.9}$$

By using (14.4) for α_2 and α_0 ($\alpha_3 \approx n\alpha_2$ and $\alpha_0 \approx n\alpha_1$) we get:

$$2(n\alpha_2 + \alpha_1 - \alpha_2 - n\alpha_1) = \frac{R_2[2(n\alpha_2 + \alpha_1 - \alpha_2) - n\alpha_1]}{f} \tag{14.10}$$

By simplifying (14.10), we finally get the "working" expressions (14.11) and (14.12), which connect focal length f, refractive index n, external R_2 and internal R_1 vial radii:

$$f = R_2 \frac{n(2R_2 - R_1) - 2(R_2 - R_1)}{2(n-1)(R_2 - R_1)} = \frac{R_2}{2} \frac{1}{n-1} \left[n \frac{2R_2 - R_1}{R_2 - R_1} - 2 \right] \tag{14.11}$$

$$n = \frac{2(R_2 - f)}{\left[\dfrac{2R_2 - R_1}{R_2 - R_1} R_2 \right] - 2f} \tag{14.12}$$

Since the focal length f and the vial radii (R_1 and R_2) can be measured directly, the refractive index n of the vial material can be calculated using (14.12).

(b) Filled Cylindrical Vial

Consider the case when the cylindrical vial is filled with transparent liquid. The refractive index of the empty cylinder is n_c (of course, $n_c \equiv n$ from the previous section), and the refractive index of the liquid is n_l. Consider a light ray falling on the filled cylindrical vial from the left, as shown in Fig. 14.6a. Direct ray tracing shows that the filled cylindrical vial acts as a converging cylindrical lens. Numerical values of angles $\alpha_0 \ldots \alpha_6$ are presented in Table 14.2, assuming that the refractive index of the vial material n_c is 1.5, and the refractive index of the liquid n_l is 1.3.

Figure 14.6 allows us to derive an expression for the focal length of the filled cylindrical vial in a similar way as done in the previous section for the case of the empty cylindrical vial.

Table 14.2 Numerical values of angles α_0 … α_6 shown in Fig. 14.6 (refractive indices n_c and n_l are assumed to be 1.5 and 1.3, respectively)

α_0	$32°$
α_1	$21°$
α_2	$31°$
α_3	$36°$
α_4	$36°$
α_5	$31°$
α_6	$21°$
α_7	$32°$

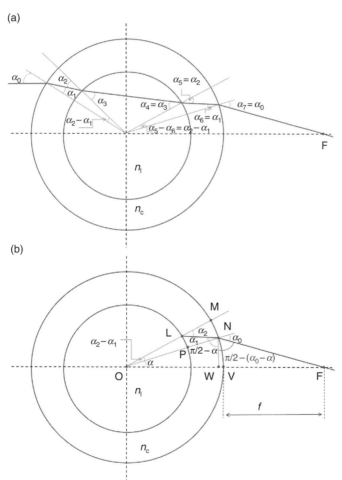

Figure 14.6 Filled cylindrical vial acts as a converging cylindrical lens.

$$\alpha = \pi - [\alpha_0 + 2(\alpha_2 - \alpha_1) + (\pi - 2\alpha_3)] = 2(\alpha_3 + \alpha_1 - \alpha_2) - \alpha_0 \quad (14.13)$$

$$\sin(\alpha) = \frac{NW}{ON} = \frac{NW}{R_2} \Rightarrow NW = R_2 \sin\alpha \quad (14.14)$$

$$WV = OV - OW = R_2 - R_2 \cos\alpha \quad (14.15)$$

$$\tan(\alpha_0 - \alpha) = \frac{NW}{WF} = \frac{R_2 \sin\alpha}{f + WV} = \frac{R_2 \sin\alpha}{f + R_2(1 - \cos\alpha)} \quad (14.16)$$

In addition to expressions (14.13)–(14.16), we use Snell's law $\frac{\sin(\alpha_0)}{\sin(\alpha_1)} = n_c$ $\frac{\sin(\alpha_3)}{\sin(\alpha_2)} = \frac{n_c}{n_l}$, and the already known expression (14.5): $\frac{\alpha_2}{\alpha_1} = \frac{R_2}{R_1}$.

Again, let us consider the paraxial approximation, where $\tan\alpha \approx \sin\alpha \approx \alpha$; $\cos\alpha \approx 1$. In this case, expression (14.16) is simplified to the following form:

$$\alpha_0 - \alpha = \frac{R_2 \alpha}{f} \Rightarrow f = \frac{R_2 \alpha}{\alpha_0 - \alpha} \quad (14.17)$$

Snell's law allows us to simplify expression (14.13):

$$\alpha_0 - \alpha = 2\alpha_0 - 2(\alpha_3 + \alpha_1 - \alpha_2) = 2(\alpha_0 + \alpha_2 - \alpha_3 - \alpha_1)$$

$$= 2\left(\alpha_1 n_c + \alpha_2 - \alpha_2 \frac{n_c}{n_l} - \alpha_1\right) \rightarrow$$

$$\alpha_0 - \alpha = 2\left[\alpha_1(n_c - 1) - \alpha_2\left(\frac{n_c}{n_l} - 1\right)\right] \quad (14.18)$$

$$\alpha = 2\left(\alpha_2 \frac{n_c}{n_l} + \alpha_1 - \alpha_2\right) - \alpha_1 n_c = \alpha_2 2\left(\frac{n_c}{n_l} - 1\right) + \alpha_1(2 - n_c) \quad (14.19)$$

Now substitute expressions (14.18) and (14.19) for α and $\alpha_0 - \alpha$ into (14.17), to get final "working" expressions for focal length f, refractive index of liquid n_l, and refractive index n_c of the material of the cylindrical vial:

$$f = \frac{R_2\left[\alpha_2 2\left(\frac{n_c}{n_l} - 1\right) + \alpha_1(2 - n_c)\right]}{2\left[\alpha_1(n_c - 1) - \alpha_2\left(\frac{n_c}{n_l} - 1\right)\right]} = \frac{R_2}{2} \frac{2\frac{R_2}{R_1}\left(\frac{n_c}{n_l} - 1\right) + 2 - n_c}{n_c - 1 - \frac{R_2}{R_1}\left(\frac{n_c}{n_l} - 1\right)} \quad (14.20)$$

$$n_l = \frac{\dfrac{2n_c(R_2 + f)}{R_1}}{n_c\left[\dfrac{2f}{R_2} + 1\right] + 2\left[\dfrac{f}{R_1} - \dfrac{f}{R_2} + \dfrac{R_2}{R_1} - 1\right]} = \frac{2n_c(R_2 + f)}{2(R_2 - R_1)\left[\dfrac{f}{R_2} + 1\right] + R_1 n_c\left[\dfrac{2f}{R_2} + 1\right]} \quad (14.21)$$

$$n_c = \cfrac{2\cfrac{R_2}{R_1} - \cfrac{R_1 f + R_2}{R_2}}{\left[\cfrac{2}{n_l}\cfrac{(f + R_2)}{R_1} - \cfrac{2f + R_2}{R_2}\right]} = \frac{2n_l(R_2 - R_1)(R_2 + f)}{2R_2(f + R_2) - R_1 n_l(R_2 + 2f)} \qquad (14.22)$$

Check expression (14.20): If $n_c = n$ and $n_l = 1$, we should get (14.11):

$$f = \cfrac{R_2}{2}\cfrac{2\cfrac{R_2}{R_1}(n - 1) + 2 - n}{n - 1 - \cfrac{R_2}{R_1}(n - 1)}$$

$$= -\frac{R_2}{2}\frac{1}{n - 1}\left[n\frac{2R_2 - R_1}{R_2 - R_1} - 2\right] \qquad (14.23)$$

As can be seen, expression (14.23) equals expression (14.11) taken with opposite sign; the minus sign appears because, in the case of the empty cylindrical vial, the back focal point is virtual. The filled cylindrical vial has a real back focal point.

Similarly, expression (14.22) transforms to the expression (14.12) when we put $n_l = 1$ and $f = -f$.

In all the expressions (14.1)–(14.22), *we use absolute (positive) values* of physical quantities (they can be modified by using sign conventions; interested readers can do this by themselves).

Materials Needed

- Optical table or optical rail
- Light sources (red cw laser, tungsten bulb)
- Object (arrow or cross)
- Ruler
- Mechanical holders
- Screws and screwdrivers
- Calipers and/or spherometer
- Squared graph paper
- Reticle
- Diaphragm
- Slit
- Screen
- Needles
- Mirror
- Spherical and cylindrical lenses: plano-concave, plano-convex
- Empty cylindrical vial
- Liquid (distilled water)

Procedures

General Properties of Cylindrical Lenses

1. Identify a converging cylindrical lens available in your optics kit.

2. Place the cylindrical lens between a rectangular slit and a screen in such a way that the axis of the cylinder is parallel to the longest side of the slit. Illuminate the slit with a collimated light beam (previous chapters discuss making a collimated light beam). Observe the light passing through the lens on the screen, as shown in Fig. 14.7a.

3. Rotate the cylindrical lens so the axis of the cylinder is perpendicular to the longest side of the slit (Fig. 14.7b). Observe

(a)

(b)

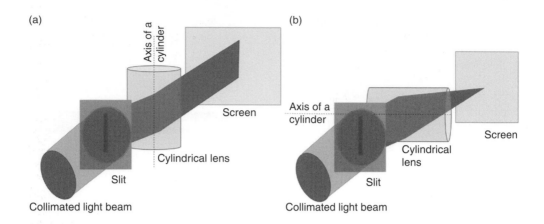

Axis of a cylinder

Screen

Cylindrical lens

Slit

Collimated light beam

Axis of a cylinder

Screen

Cylindrical lens

Slit

Collimated light beam

Figure 14.7 Focusing properties of the converging cylindrical lens: (a) Axis of cylinder is parallel to longest side of slit; (b) axis of cylinder is perpendicular to longest side of slit.

the light passing through the lens on the screen. (Hint: move the screen to bring the light beam into focus.)

4. Replace the converging cylindrical lens with a diverging cylindrical lens and repeat steps 2 and 3.

5. Draw a conclusion about the focusing properties of the cylindrical lenses.

Characterization of Cylindrical Lenses

(a) "At-a-Glance" Characterization

As in the case of spherical lenses (for details, see Chapter 6), a similar express method, the so-called "at-a-glance" technique, can be used to make basic characterization of cylindrical lenses.

The type of a cylindrical lens (thin or thick, plano-convex, plano-concave, double-concave, double-convex, concave-convex, meniscus) can be determined by visual observation of the lens shape. In addition, visual analysis allows us to determine the general optical quality of the lens (absence or presence of any defects such as air bubbles, scratches etc.). The focal length of the cylindrical lens can be determined using the simple scheme shown in Fig. 14.8, according to the following procedure:

1. Place a cylindrical lens on a squared paper sheet so that the axis of the cylinder is parallel to the grid line (y-axis in Fig. 14.8). Notice the visible change in the size of the square grid by comparing areas of paper covered and not covered by the lens. The grid period d_x along the x-axis is changed, but the grid period along the y-axis stays the same (Fig. 14.8).

Cylindrical Lenses and Vials

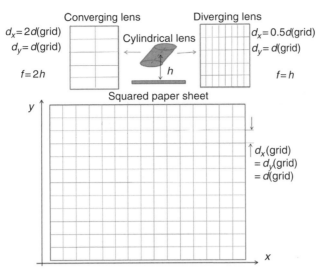

Figure 14.8
"At-a-glance" characterization of a cylindrical lens.

2. Take the lens and raise it to increase the distance *h* between the paper's surface and the lens. Simultaneously watch the changes in size d_x of the square grid observed with the lens. Continue to increase the distance between the lens and paper sheet until the observed period d_x of the square grid is double-changed (increased by a factor of 2 for a converging cylindrical lens and decreased by a factor of 2 for a diverging cylindrical lens, as shown in Fig. 14.8). Measure the distance *h* between the paper surface and the lens found for each lens. As can be shown via direct geometrical ray tracing, such a distance *h* is related to the focal length *f* of the cylindrical lenses by simple expressions (derive them): $f = 2h$ for the converging cylindrical lens, and $f = h$ for the diverging cylindrical lens.

3. Complete the following table.

   Table 14.3

Lens	Focal length* (mm)	Refracting power (D)	Lens size (mm)
#1			
#2			
#3			
#4			
...			

   * Remember the sign conventions: Focal length is positive for a converging lens and negative for a diverging lens.

4. Estimate the time you need to complete an "at-a-glance" characterization of one cylindrical lens.

Figure 14.9 Optical set-up to measure the focal length of a thin diverging lens using a laser beam.

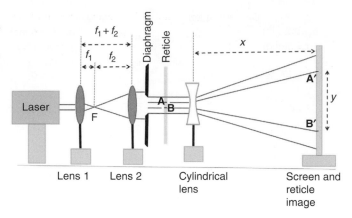

(b) Diverging Cylindrical Thin Lens

Use an expanded laser beam as a parallel light beam. A real laser beam has a finite angular divergence. To produce a good result, the angular divergence of the output light beam, which is caused by the diverging lens, should be much greater than the intrinsic angular divergence of the laser beam.

1. Identify two spherical thin lenses (lens 1 and lens 2) and a thin diverging cylindrical lens available in your optics kit. Assemble the optical set-up shown in Fig. 14.9.

2. Place a screen at eight different positions with respect to the diverging lens. For each position, measure the distance x_i and the width of the transmitted laser beam y_i. Complete the following table.

   Table 14.4

x_i (mm)								
y_i (mm)								

3. Plot the dependence of y_i on x_i. As shown directly in Figs. 14.9 and 14.10, the expected dependence is $y = \text{const}(x + f)$.

4. By fitting the plotted dependence with a straight line, determine the focal length of a diverging cylindrical lens, as shown in Fig. 14.11.

5. Write down the results obtained for a diverging cylindrical lens:
   $f =$

6. Explain why we use a diaphragm and a reticle in the experimental set-up. Draw a conclusion about the limitations of the considered experimental method.

Cylindrical Lenses and Vials

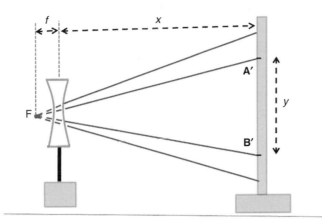

Figure 14.10 Focal length of a diverging cylindrical lens.

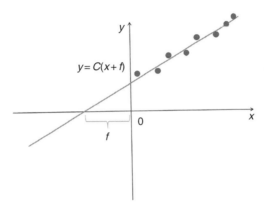

Figure 14.11 Determination of the focal length by fitting experimental data with a linear dependence.

(c) Converging Cylindrical Thin Lens

1. Identify two spherical thin lenses (lens 1 and lens 2) and a thin converging cylindrical lens available in your optics kit. Assemble the optical set-up shown in Fig. 14.12.

2. Place the screen at eight different positions with respect to the converging lens. For each position, measure the distance x_i and the width of the transmitted laser beam y_i. Complete the following table.

Table 14.5

x_i (mm)								
y_i (mm)								

Figure 14.12 Optical set-up to measure the focal length of a thin converging lens by using a laser beam.

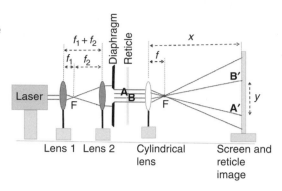

Figure 14.13 Determination of the focal length by fitting experimental data with a linear dependence.

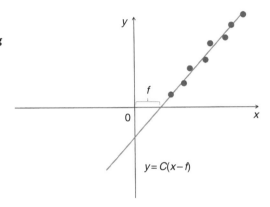

3. Plot the dependence of y_i on x_i. As shown directly in Fig. 14.12, the expected dependence is $y = \text{const}(x - f)$.

4. Fit the plotted dependence with a straight line to determine the focal length of the converging cylindrical lens, as shown in Fig. 14.13.

5. Write down the results obtained for a converging cylindrical lens:
   $f =$

6. Draw a conclusion about the limitations of the considered experimental method.

(d) Converging or Diverging Cylindrical Thick Lenses

The experimental procedures described above are also valid for the case of *thick* cylindrical lenses. All the arrangements are the same (source of collimated light, laser and beam expander, diaphragm, reticle, cylindrical

(a) (b)

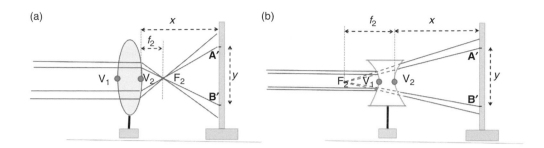

thick lens, screen). However, a difference is that distance "*x*" is measured between the back vertex and a screen, as shown in Fig. 14.14 (for the case when falling collimated light is coming from the left to the right).

Figure 14.14 Back focal points (F₂) and back focal lengths (*f₂*) of thick cylindrical lenses: (a) Converging cylindrical lens; (b) diverging cylindrical lens.

1. Identify thick converging and diverging cylindrical lenses available in your optics kit.

2. Use the procedures (b) and (c) described above to measure the back focal length f_2 of thick cylindrical lenses. Place the measured values in the following table.

Table 14.6

x_i (mm)							
y_i (mm)							

3. Make the collimated light beam fall on the lenses from the right to the left. Similarly, measure the front focal length of the lenses. Place the measured values in the following table.

Table 14.7

x_i (mm)							
y_i (mm)							

4. Complete the following table.

Table 14.8

Cylindrical lens	Front focal length, f_1 (mm)	Back focal length, f_2 (mm)
Converging		
Diverging		

189

Determination of the Refractive Index of Liquids Using Cylindrical Vials

A cylindrical vial can be used to determine both the refractive index of the vial material and the refractive indices of unknown liquids.

(a) Empty Cylindrical Vial Acts As a Diverging Cylindrical Lens

1. Identify spherical lenses (to make a collimated light beam) and a cylindrical vial, available in your optics kit. Assemble the optical set-up shown in Fig. 14.15. Place a diaphragm before the reticle to justify the use of the paraxial approximation.

2. Measure the internal R_1 and external R_2 radii of the cylindrical vial using calipers.

3. Place a screen at eight different positions with respect to the cylindrical vial. For each position, measure the distance x_i and the width of the transmitted light beam y_i. Measure the distance x as shown in Fig. 14.16. You cannot treat a cylindrical vial as thin, so it is important to measure the correct distance.

Figure 14.15 Optical set-up to determine the refractive index of the cylindrical vial material.

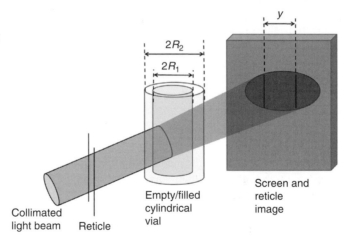

Figure 14.16 Explanation of the meaning of the quantity x.

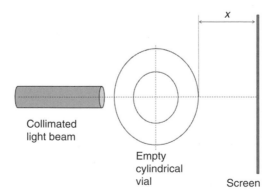

4. Complete the following table.

Table 14.9

x_i (mm)							
y_i (mm)							

5. Plot the dependence of y_i on x_i, and, by fitting the plotted dependence with a straight line $y = \text{const}(x + f)$, determine the focal length f of the cylindrical vial as in the previous procedures.

6. Using an expression derived in the Background section, calculate the refractive index of the vial material. Complete the following table.

Table 14.10

R_1 (mm)	R_2 (mm)	f (mm)	n^*

$$^* n = \frac{2(R_2 - f)}{\left[\dfrac{2R_2 - R_1}{R_2 - R_1} R_2\right] - 2f}$$

7. Estimate the experimental errors and draw a conclusion about the limitations of the considered experimental method.

(b) Cylindrical Vial Filled with Liquid Acting As a Converging Cylindrical Lens

1. Fill a cylindrical vial with the liquid to be studied (use the same optical set-up shown in Fig. 14.15 of the previous procedure).

2. Place a screen at eight different positions with respect to the filled cylindrical vial. For each position, measure the distance x_i and width of the transmitted light beam y_i. Measure the correct distance x as shown in Fig. 14.17.

3. Complete the following table.

Table 14.11

x_i (mm)							
y_i (mm)							

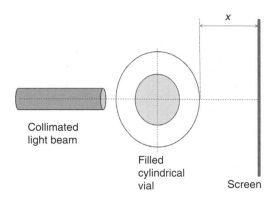

Collimated
light beam

Filled
cylindrical
vial

Screen

4. Plot the dependence of y_i on x_i. Fit the plotted dependence with a straight line $y = \text{const}(x - f)$ and determine the focal length f of the filled cylindrical vial as in previous procedures.

5. By using an expression derived in the Background section, calculate the refractive index of the liquid. Complete the following table.

Table 14.12

R_1 (mm)	R_2 (mm)	f (mm)	$n_c{}^*$	$n_l{}^{**}$

* $n_c = n$ (found in a previous procedure)

$$^{**}\ n_l = \frac{\dfrac{2n_c(R_2+f)}{R_1}}{n_c\left[\dfrac{2f}{R_2}+1\right] + 2\left[\dfrac{f}{R_1} - \dfrac{f}{R_2} + \dfrac{R_2}{R_1} - 1\right]} = \frac{2n_c(R_2+f)}{2(R_2 - R_1)\left[\dfrac{f}{R_2}+1\right] + R_1 n_c\left[\dfrac{2f}{R_2}+1\right]}$$

6. Estimate the experimental errors and draw a conclusion about the limitations of the considered experimental method.

Spherical Aberrations of Converging Cylindrical Lenses

1. Identify converging cylindrical lenses available in your optics kit.

2. Assemble a laser beam expander, and place a Hartmann screen behind the expander as shown in Fig. 14.18.

3. Place the test cylindrical lens behind the Hartmann screen as shown in Fig. 14.18. Use a cardboard screen to locate the

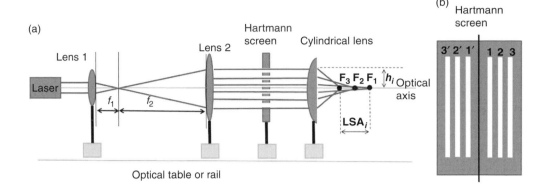

position of the back focal point for each pair of rays propagating at a distance h_i relative to the optical axis. Hint: To locate the back focal points F_i, use one pair of rays for each focal point. For example, if the Hartmann screen looks like the one shown in Fig. 14.18b, rays coming from the pair of holes/slits 1 and 1' will converge into back focal point F_1. The other back focal points can be found analogously. During determination of a particular back focal point, block the rest of the light rays with a blocking screen.

Figure 14.18
(a) Optical set-up to measure spherical aberration of a cylindrical lens using a Hartmann screen (b).

4. Locate the paraxial back focal point of the lens. Measure LSA (the distance between the real back focal point and the paraxial focal point) for each light ray propagating at a distance h_i from the optical axis, as shown in Fig. 14.18. Complete the following table.

Table 14.13

h (mm)					
LSA (mm)					

5. Plot the dependence of LSA on h^2.

6. Draw a conclusion about the qualitative value of the spherical aberration and its dependence on falling ray parameters.

Evaluation and Review Questions

1. Analyze the experimental methods described in Chapter 6 to characterize thin spherical lenses. Develop similar experimental procedures to measure focal lengths of cylindrical lenses.

2. How can one determine the refractive index of the material of a cylindrical lens? Develop appropriate experimental procedures. (Hint: Use the expression which connects the focal length of the lens, its radius of curvature, and its refractive index.)
3. Explain why a cylindrical lens can correct astigmatism.
4. By using any available math software (Mathematica, MathCad, MatLab, Origin, Maple, etc.), plot the dependence of the focal length of the cylindrical vial versus a ratio of the external and internal radii.
5. Draw an exact ray tracing for the case when the cylindrical vial is filled with a material whose refractive index n_m is larger than the refractive index n_c of the cylindrical vial: $n_c < n_m$.

Further Reading

General

E. Hecht, *Optics*, 4th edition, San Francisco, CA: Addison-Wesley, 2001

F. L. Pedrotti, S. J. L. Pedrotti, L. M. Pedrotti, *Introduction to Optics*, 3rd edition, Upper Saddle River, NJ: Pearson Prentice Hall, 2007

Specialized

Handbook of Optics, W. G. Driscoll (editor), W. Vaughan (associate editor), New York: McGraw-Hill, 1978

D. C. O'Shea, *Elements of Modern Optical Design*, New York: Wiley, 1985

Chapters 2, 5, and 6 of this book

Methods of Geometrical Optics to Measure Refractive Index

15

Objectives

1. Measure the refractive indices of solids and liquids by different methods:

   - Direct ray tracing
   - Traveling microscope (Duc de Chaulnes' image displacement method)
   - Methods of total internal reflection or critical angle measurement (also by use of refractometers – Abbe, Pfund, Pulfrich)
   - Methods of minimum angle deviation
   - Method of liquid lenses (by using both cylindrical and spherical lenses)
   - Laser beam displacer
   - Method of hollow cells (both rectangular and prismatic)
   - Method of liquid immersion (including refractive index-matching liquids)

2. Compare these techniques for experimental errors, time consumption, affordability, reliability, amount of materials needed to complete measurements, and universality.

Background

Direct Ray Tracing

The simplest way to measure the refractive index of materials is direct ray tracing. This process includes tracing an input/output light beam, measuring angles of incidence α_i and of refraction α_r. Refractive index n can be found by applying Snell's law ($\sin \alpha_o / \sin \alpha_r = n$). This method was described in Chapter 5. The drawbacks of this method include low accuracy and special material requirements, such as an exact geometrical shape.

Recently, an interesting modification of direct ray tracing has been proposed in several papers (e.g. S. Lombardi, G. Monroy, I. Testa, E. Sassi, Measuring variable refractive indices using digital photos, *Physics Education*, **45**(1), 83–92 (2010)). The authors proposed taking a digital picture of the laser rays propagating through a medium and measuring all the required angles from this photo.

Traveling Microscope (Duc de Chaulnes' Image Displacement Method)

If an object S is observed through a transparent slab in the shape of a parallel plate, it appears displaced along the direction of observation by a distance $SS' = \delta$, as shown in Fig. 15.1. This fact can be used to measure the refractive index of the slab. Using the paraxial approximation and Snell's law, we can write equations (15.1)–(15.4), which follow directly from Fig. 15.1:

$$\frac{x}{h} = \tan(\alpha_0) \approx \alpha_0 \tag{15.1}$$

$$\frac{y}{d} = \tan(\alpha) \approx \alpha \tag{15.2}$$

$$\frac{\alpha_0}{\alpha} = n \tag{15.3}$$

$$\frac{x+y}{d+h-\delta} = \tan(\alpha_0) \approx \alpha_0 \tag{15.4}$$

The refractive index is found by substituting equations (15.1)–(15.3) into (15.4). This leads to equation (15.5):

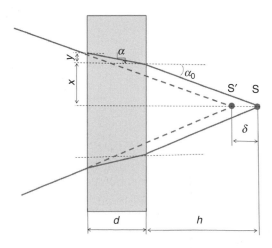

Figure 15.1 Traveling microscope method to determine refractive index.

$$n = \frac{d}{d - \delta} \tag{15.5}$$

Typically, a microscope is used to measure the refractive index of a slab. An experimentalist must focus sequentially on the front and back surfaces of a slab of known thickness d. To do that, the objective of the microscope is displaced by a distance δ. That is why this method is called the "traveling microscope." Knowing the values of d and δ, the refractive index of the slab can be found from equation (15.5).

Methods Utilizing the Phenomenon of Total Internal Reflection

The method of total internal reflection was applied in Chapter 5 to determine the refractive index of a prism. There are other modifications of this technique which are discussed briefly below.

Pfund's Method

This method can be applied to measure the refractive index of a transparent slab (Fig. 15.2a) and a thin layer of liquid covering the slab (Fig. 15.2b).

Consider a point source emitting light as shown in Fig. 15.2a. The condition of total internal reflection can be written as:

$$\sin \alpha_c = \frac{(R/2)}{\sqrt{(R/2)^2 + d^2}} = \frac{1}{n_g} \tag{15.6}$$

This equation yields an expression for the refractive index of a slab:

$$n_g = \frac{\sqrt{D^2 + 16d^2}}{D} \tag{15.7}$$

In a similar way, the refractive index of a thin liquid layer covering the slab, as shown in Fig. 15.2b, is found according to:

$$\sin \alpha_c = \frac{(R/2)}{\sqrt{(R/2)^2 + d^2}} = \frac{n_l}{n_g} \tag{15.8}$$

$$n_l = \frac{n_g D}{\sqrt{D^2 + 16d^2}} \tag{15.9}$$

Basics of Refractometers

A prism of known refractive index n_{prism} can be used to measure the refractive index n_{liquid} of a liquid sample covering the prism, as shown in Figure 15.3. By applying Snell's law, we can write:

Figure 15.2 Pfund's method to measure refractive index: (a) Determination of the refractive index n_g of the glass plate; (b) determination of the refractive index n_l of the liquid layer on the same glass plate.

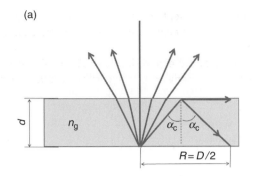

(a)

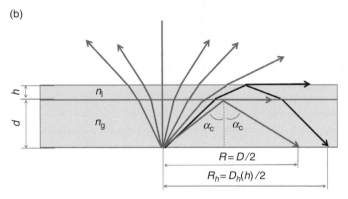

(b)

Figure 15.3 Principle of the refractometer.

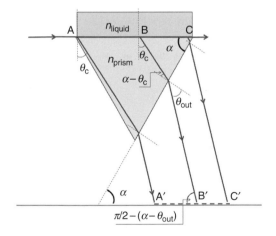

$$n_{prism} = \frac{\sin(\theta_{out})}{\sin(\alpha - \theta_c)} \qquad (15.10)$$

$$\frac{n_{liquid}}{n_{prism}} = \sin(\theta_c) \qquad (15.11)$$

Experimentally, it is more convenient to rewrite the equation for the refractive index using the refractive angle α of a prism, and the emergent angle θ_{out} shown in Fig. 15.3.

In this case, the following expression is used to calculate the refractive index of a liquid (its derivation is recommended as a good morning problem):

$$n_{liquid}^2 = \frac{n_{prism}^2}{1 + \left[\frac{\sin\theta_{out} + \cos\alpha}{\sin\alpha}\right]^2} \tag{15.12}$$

As shown in this expression, the refractive index of a liquid is described mathematically as a function of the emergent angle θ_{out}. This means that different liquids placed on top of the same prism bend emergent light quantitatively in different ways; the emergent angle is changed according to expression (15.12) above. As a result, the scale on the screen can be graduated to represent the value of the refractive index of the liquid.

Existing refractometers (the most popular among them are the Abbe, Pulfrich, and Pfund types) have different designs. They can be analog or digital, and are available for laboratory or home use, but all have one common feature. Their basic principle of operation is based on the phenomenon of total internal reflection. We do not discuss the designs of available refractometers, but refer interested readers to the device manuals, which include the best descriptions available.

Method of a Prism

If the material can be shaped in the form of a prism, its refractive index is found according to the following equations (Fig. 15.4):

$$\alpha = \theta_1' + \theta_2' \tag{15.13}$$

$$\delta = \left(\theta_1 - \theta_1'\right) + \left(\theta_2 - \theta_2'\right) = \theta_1 + \theta_2 - \alpha \tag{15.14}$$

$$n = \frac{\sin\theta_1}{\sin\theta_1'} = \frac{\sin\theta_2}{\sin\theta_2'} \tag{15.15}$$

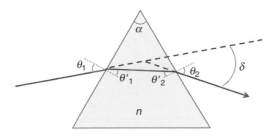

Figure 15.4 Method of a prism to measure refractive index (general case).

Method of Minimum Angle Deviation

This classical method is based on the following expression (please read any optics textbook to find out more about the derivation of this formula):

$$n = \frac{\sin\left[(\alpha + \delta_{min})/2\right]}{\sin(\alpha/2)} \qquad (15.16)$$

Method of Normal Incidence (Wolfe's Method, $\theta_1 = \pi/2$)

Assuming that light falls on a prism at right angles, as shown in Fig. 15.5, directly applying Snell's law gives us the following formula:

$$n = \frac{\sin(\theta_2)}{\sin(\theta'_2)} = \frac{\sin(\alpha + \delta)}{\sin(\alpha)} \qquad (15.17)$$

According to this expression, to determine the refractive index of a prism material, measure the refractive angle α of the prism and the angle of deviation δ. The method of normal incidence is used when the refractive angle α is smaller than the critical angle α_c of total internal reflection: $\alpha < \alpha_c$.

Method "$\theta_1 = \alpha$"

In this variation of the prism method, the angle of incidence θ_1 is equal to the refractive angle α of the prism: $\theta_1 = \alpha$. In other words, the extension of the incident ray is perpendicular to the exit surface of the prism, as shown in Fig. 15.6. As a result, the exit angle θ_2 is equal to the deviation angle δ: $\theta_2 = \delta$ (Fig. 15.6).

Apply Snell's law: $n = \sin\alpha/\sin\theta'_1 = \sin\delta/\sin\theta'_2$. Since $\alpha = \theta'_1 + \theta'_2$, rewrite the expression in the following form: $n = \sin\alpha/\sin(\alpha - \theta'_2) = \sin\delta/\sin\theta'_2$. As a result, we have a system of two equations:

$$\sin\alpha\sin\theta'_2 = \sin\delta\sin(\alpha - \theta'_2) \qquad (15.18)$$

$$n^2 = \frac{\sin^2\delta}{\sin^2\theta'_2} \qquad (15.19)$$

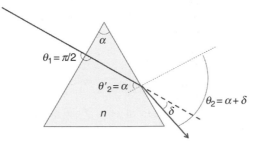

Figure 15.5 Method of normal incidence to measure refractive index.

$\theta_1 = \pi/2$

α

$\theta'_2 = \alpha$

n

δ

$\theta_2 = \alpha + \delta$

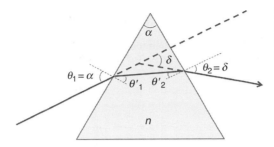

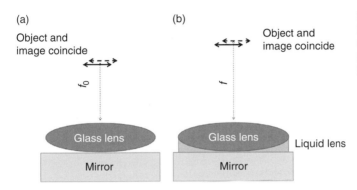

(a)

Object and
image coincide

(b)

Object and
image coincide

Figure 15.7 Method of liquid lenses to determine the refractive index of a liquid.

Once $\sin \theta'_2$ has been found from the first expression, and the result substituted into the second expression, the final formula has the following form:

$$n^2 = \sin^2 \delta + [1 + \sin \delta \operatorname{ctg} \alpha]^2 \qquad (15.20)$$

Method of Liquid Lenses

(a) Spherical Lenses

This technique uses the lensmaker's equation: $\frac{1}{f} = \left(\frac{n-n_0}{n}\right)\left(\frac{1}{R_1} - \frac{1}{R_2}\right)$, and the common fact that a liquid takes the shape of its container.

     Consider a thin glass lens placed in close contact on a mirror, as shown in Fig. 15.7a. If an object is placed at a distance which equals the focal length f_0 of the glass lens, light transmitted through the lens will be reflected by the mirror. Reflected light forms an image which coincides with the object. If a few drops of liquid are placed between the glass lens and the mirror, as shown in Fig. 15.7b, then we have a combination of two lenses: a converging glass lens and diverging liquid lens. These lenses are thin and in close contact, and their new focal length f can be found as described for the previous case. We can write the following lensmaker's expressions for cases in Fig. 15.7a and b. Here R_1 and R_2 are the radii of curvature of the glass lens, f_{liquid} is the focal length of the liquid lens, and n_{liquid} is the refractive index of the liquid:

$$\frac{1}{f_0} = \left(n_{glass} - 1\right)\left(\frac{1}{R_1} - \frac{1}{R_2}\right) \tag{15.21}$$

$$\frac{1}{f_{liquid}} = \left(n_{liquid} - 1\right)\left(\frac{1}{R_2} - \frac{1}{\infty}\right) \tag{15.22}$$

$$\frac{1}{f} = \frac{1}{f_0} + \frac{1}{f_{liquid}} \tag{15.23}$$

$$\frac{1}{f} = \left(n_{glass} - 1\right)\left(\frac{1}{R_1} - \frac{1}{R_2}\right) + \left(n_{liquid} - 1\right)\frac{1}{R_2}$$
$$= \left(n_{glass} - 1\right)\frac{1}{R_1} + \frac{1}{R_2}\left(n_{liquid} - n_{glass}\right) \tag{15.24}$$

These expressions are simplified in the case of the plano-convex glass lens ($R_1 = \infty$):

$$\frac{1}{f_0} = -\left(n_{glass} - 1\right)\frac{1}{R_2} \tag{15.25}$$

$$\frac{1}{f_{liquid}} = \left(n_{liquid} - 1\right)\frac{1}{R_2} \tag{15.26}$$

$$\frac{1}{f} = \frac{1}{f_0} + \frac{1}{f_{liquid}} = \frac{1}{R_2}\left(n_{liquid} - n_{glass}\right) \tag{15.27}$$

Thus, the refractive index of the liquid n_{liquid} can be expressed in terms of the physical quantities (R_2, f, f_0) which are measured directly:

$$n_{liquid} = \frac{R_2}{f_{liquid}} + 1 = \left(\frac{1}{f} - \frac{1}{f_0}\right)R_2 + 1 \tag{15.28}$$

(b) Cylindrical Lenses

This case was discussed in Chapter 14.

Laser Beam Displacer

This method was described in Chapter 5.

Method of Hollow Cells

When hollow cells (prisms, plane plates) are filled with liquid, the refractive index of the liquid can be determined similarly to the cases already described (methods of the prism, laser beam displacer, traveling microscope). When the walls of the hollow cell are very thin, their effect on light beam bending can be neglected, and the filled cell can be treated as a liquid cell. This uses all the appropriate equations shown above. However, precise measurements of the refractive index require special consideration of the effect of the cell wall thickness on the measured values. We do not discuss these topics here, referring interested readers to the special literature. Note that, in the case when the thickness of the cell wall is taken into consideration, the working formulas are a bit complicated.

Method of Liquid Immersion (Using Refractive Index-Matching Liquids)

In the method of liquid immersion, the object (powder, piece of glass, lens, biological tissue, etc.) is immersed in a homogeneous liquid (the so-called immersion liquid) whose index can be varied. The liquid is a mixture of two miscible compounds, one whose index is higher than the object index and one whose index is lower. By varying the concentration of these components, the refractive index of the immersion liquid can match the refractive index of the object. A table of suitable liquids can be found in the *CRC Handbook of Chemistry and Physics*.

There are two basic schemes to determine refractive index using the liquid immersion method: the Becke line effect and the collimated target method.

The Becke line effect

This scheme of measurement requires using a microscope, as shown in Fig. 15.8. If the refractive index of the object (a powder in Fig. 15.8) is larger (Fig. 15.8a) or smaller (Fig. 15.8b) than the index of the immersion liquid, the object is clearly visible. In these cases, (a) and (b) in Fig. 15.8, the Becke line appears as a band of light along the boundary between powder and liquid. To observe the Becke line, the objective of the microscope

Figure 15.8 Liquid immersion method to measure the refractive index of powders immersed in a liquid (using the Becke line effect).

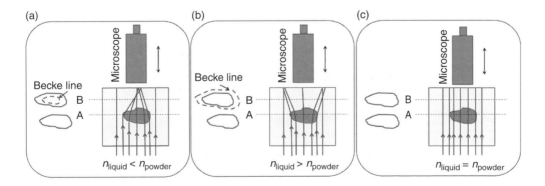

should be focused on the object plane A, and then at plane B which is very close to plane A, as shown in Fig. 15.8. The size of the Becke line directly indicates the difference in refractive index between the powder and the immersion liquid. It allows us to adjust the refractive index of the liquid to match the refractive index of the powder. When the refractive indices of the powder and the liquid are equal, the Becke line disappears, and the object becomes "invisible" (i.e. we do not see sharp borders between the object and the liquid).

Materials Needed

- Optical table or optical rail
- Light sources (red cw laser, tungsten bulb)
- Object (arrow or cross)
- Ruler
- Mechanical holders
- Screws and screwdrivers
- Calipers and/or spherometer
- Squared graph paper
- Screen
- Needles
- Mirror
- Thin lenses: plano-convex
- Diaphragm (a simple opening made of paper may suffice) and adhesive tape to fix the diaphragm to a holder
- Reticle
- Liquid (water)
- Pipette
- Piece of glass 1 inch × 1 inch
- Spacer (Teflon strip, adhesive tape)
- Screen and cardboard

Collimated target scheme

In this scheme of measurement, a collimated target is viewed through the "powders immersed in liquid" system by using a telescope focused on infinity. When the refractive index of the immersion liquid does not match the refractive index of the powder, the light beam leaving the plane liquid cell is converging or diverging. This depends on the sign of the difference in refractive index between the liquid and the powder. If the refractive indices are mismatched, the observed clarity of the target is very low. Only when the refractive indices of the immersion liquid and the powder match does the emergent light beam remain collimated. In addition, the clarity of the observed target is very sharp.

Procedures

Wear gloves to handle optical elements (lenses).

Prototype of the refractometer

1. Identify the equal angle ($\alpha = 60°$) prism available in your optics kit.

2. Assemble the following optical set-up, as shown in Fig. 15.9a. Direct a laser beam at a grazing angle (along the top surface of the prism) and observe the refracted light pattern on the screen.

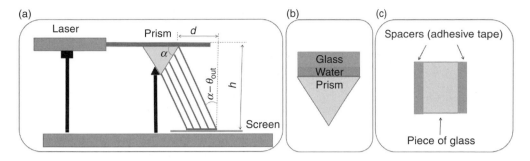

Figure 15.9
Experimental set-up.

3. Measure distances d and h as shown in Fig. 15.9a. Calculate $\sin(\alpha - \theta_{out}) = \frac{d}{\sqrt{d^2 + h^2}}$, then find the emergent angle θ_{out}, and determine the refractive index of the prism: $n_{prism}^2 = 1 + \left[\frac{\sin\theta_{out} + \cos\alpha}{\sin\alpha}\right]^2$. Complete the following table.

Table 15.1

d (mm)	h (mm)	$\sin(\alpha - \theta_{out})$	$\alpha - \theta_{out}$	α	θ_{out}	n_{prism}

4. Use a pipette to place a few drops of liquid (distilled water) on the top surface of the prism. On the liquid surface, place a piece of glass with two spacers (Fig. 15.9b, c). Capillary forces hold the piece of glass and prism together.

5. Adjust the laser beam (if necessary) and observe the patterns of the refracted light rays on the screen, as shown in Fig. 15.9a. Note the new positions of the refracted rays on the screen. Measure the new distance d, calculate the refractive index of the liquid, and complete the following table.

Table 15.2

d (mm)	h (mm)	$\sin(\alpha - \theta_{out})$	$\alpha - \theta_{out}$	α	θ_{out}	n_{liquid}^*

$$^* n_{liquid}^2 = \frac{n_{prism}^2}{1 + \left[\frac{\sin\theta_{out} + \cos\alpha}{\sin\alpha}\right]^2}$$

6. Think about how to graduate a screen in units of the refractive index. Estimate the experimental errors.

7. Draw a conclusion about the limitations of this experimental method.

Method of a Liquid Spherical Lens

1. Identify a plano-convex lens and mirror available in your optics kit.

2. Use a spherometer or calipers to measure the radii of curvature R_1 and R_2 of the lens. If you do not remember how to do this, see Chapter 7, procedure "Determination of the refractive index of a thick lens" for details. Remember, in the case of the plano-convex lens, $R_1 = \infty$.

3. Assemble the optical set-up shown in Fig. 15.10.

4. Use a reticle as the object. Place the reticle very close to the plane surface of the glass lens and move the reticle up. Simultaneously view the reticle image with your eye. Bring this image into sharp focus by finding the appropriate distance. Measure this distance. It should be the focal length f_0 of the glass lens.

Figure 15.10
Experimental set-up.

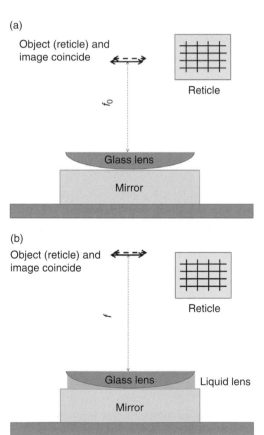

(a)

Object (reticle) and image coincide

Reticle

f_0

Glass lens

Mirror

(b)

Object (reticle) and image coincide

Reticle

f

Glass lens   Liquid lens

Mirror

5. Use a pipette and place a few drops of liquid (water) into the space between the lens and the mirror (make sure that all the space between the lens and the mirror is filled with water). Then repeat step 4 for the combined system "liquid lens–glass lens" and find a new focal length f (Fig. 15.10b).

6. Calculate the refractive index of the liquid
   $n_{liquid} = \frac{R_2}{f_{liquid}} + 1 = \left(\frac{1}{f} - \frac{1}{f_o}\right)R_2 + 1$, and complete the following table.

   Table 15.3

f_o (mm)	f (mm)	R_2 (mm)	n_{liquid}

7. Draw a conclusion about the limitations of this experimental method and discuss the factors which affect the accuracy of the results.

Evaluation and Review Question

Derive all the equations shown in the Background section.

Conclusions

Complete the following table.

Table 15.4

Refractive index of prism, n_{prism}		Refractive index of liquid, n_{liquid}		
Average value	Error	Average value	Error	Method used

Compare the studied techniques in terms of experimental errors, time consumption, fund investment, reliability, amount of materials needed to complete measurements, and universality.

For Further Investigation

By now, you are well prepared to work independently. Using information provided in the Background section, develop your own experimental procedure to measure refractive index. Consider the following methods:

- Traveling microscope (Duc de Chaulnes' image displacement method)
- Pfund's method
- Method of minimum angle deviation
- Method of normal incidence (Wolfe's method)
- Method "$\theta_1 = \alpha$"
- Method of hollow cells
- Method of liquid immersion (collimated target scheme)

Further Reading

General

E. Hecht, *Optics*, 4th edition, San Francisco, CA: Addison-Wesley, 2001

F. L. Pedrotti, S. J. L. Pedrotti, L. M. Pedrotti, *Introduction to Optics*, 3rd edition, Upper Saddle River, NJ: Pearson Prentice Hall, 2007

J. Strong, *Concepts of Classical Optics*, New York: Dover, 2004 (Dover edition is an unabridged republication of the work originally published in 1958 by W. H. Freeman and Company, San Francisco, CA)

Specialized

C. H. Palmer, *Optics: Experiments and Demonstrations*, Baltimore, MD: Johns Hopkins Press, 1962

S. Singh, Refractive index measurement and its applications, *Physica Scripta*, **65**, 167, 2002

A. F. Wagner, *Experimental Optics*, New York: John Wiley & Sons, 1929

Chapters 5 and 6 of this book

Dispersion of Light and Prism Spectroscope

16

Background

The refractive index n depends on the wavelength λ of light; this dependence is called *dispersion*. More generally, dispersion can be defined as $\partial n/\partial \lambda$. Transparent (non-absorbing) materials exhibit *normal dispersion*. Normal dispersion is when $\partial n/\partial \lambda < 0$. Absorbing materials exhibit *anomalous dispersion*, when $\partial n/\partial \lambda > 0$.

A typical normal dispersion curve is shown in Fig. 16.1.

Augustin Cauchy introduced an empirical relation that approximates the normal dispersion curve:

$$n(\lambda) = A + \frac{B}{\lambda^2} + \frac{C}{\lambda^4} + \cdots$$

where A, B, C ... are empirical constants. Very often the first two terms are sufficient to provide a reasonable fit: $n(\lambda) \approx A + B/\lambda^2$, and $\partial n/\partial \lambda = -2B/\lambda^3$.

**Figure 16.1 Typical
normal dispersion
of light.**

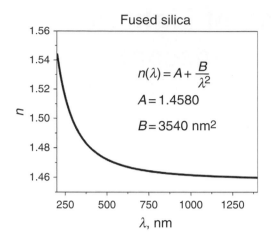

Fused silica

$$n(\lambda) = A + \frac{B}{\lambda^2}$$

$$A = 1.4580$$

$$B = 3540 \text{ nm}^2$$

λ, nm

**Figure 16.2 Deviation
and dispersion of
light propagating
through a prism.**

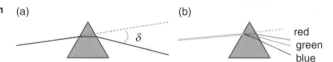

(a)

δ

(b)

red
green
blue

The *Abbe number* is widely used to describe the dispersive properties of optical materials. The Abbe number is defined as $\frac{n_D - 1}{n_F - n_C}$, where subscripts indicate certain wavelengths (F for blue light, $\lambda = 486.1$ nm; D for yellow light, $\lambda = 589.2$ nm; C for red light, $\lambda = 656.3$ nm).

When monochromatic light falls on a prism, it will only deviate, as shown in Fig. 16.2a. Under the same conditions, white light will both deviate and disperse, as shown in Fig. 16.2b.

An optical instrument employing a prism as the dispersive element, combined with the means of measuring the prism angle and the angles of deviation of various wavelength components in the incident light, is called a *prism spectrometer*. When an instrument is used for visual observations without the capability of measuring the angular displacement of the spectral "lines," it is called a *spectroscope*. If the means are provided for recording the spectrum with a photographic film or a CCD camera in the focal plane of the telescope objective, the instrument is called a *spectrograph*.

The basic components of the prism spectrometer are shown in Fig. 16.3.

A detailed description of the spectrometer is given in the Procedures section below.

Dispersion of Light and Prism Spectroscope

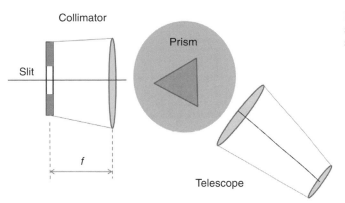

Collimator

Prism

Slit

f

Telescope

Figure 16.3 Optical scheme of the spectrometer.

Procedures

Dispersion of Light by a Prism

(a) Incandescent Bulb as Light Source

1. Identify a plano-convex lens (focal length ~100 mm, aperture diameter ~25 mm), a slit, a tungsten bulb, and an equal-angle prism in your optics kit. Assemble an optical set-up as shown in Fig. 16.4.

2. Simply connect the incandescent lamp to the DC outlet "5 V 5 A" (Fig. 16.5). Make the distance between the bulb and the lens equal to the focal length f. As a result, light emerging from the lens will be satisfactorily collimated.

3. Place a slit very close to the lens, as shown in Fig. 16.4.

4. Adjust the prism position to observe the splitting of light into a rainbow on the screen. Make sure that conditions of total internal reflection *are not satisfied*; otherwise you cannot observe any dispersion of light. Light falling on the prism should be almost perpendicular to the prism base, as shown in Fig. 16.6.

5. Write down the sequence of colors you observe on the screen.

Materials Needed

- Optical table or optical rail
- Light sources (red and green cw laser, tungsten bulb, and light emitting diode)
- Ruler
- Mechanical holders
- Screws and screwdrivers
- Calipers
- Squared graph paper
- Screen and screen with a hole
- Needles (pins)
- Thin and thick lenses
- Diaphragm (a simple opening made of paper may suffice) and adhesive tape to fix the diaphragm to a holder
- Slit
- Resistors (10 and 50 Ω) or resistor box
- Prisms
- Filters (neutral and color)

Figure 16.4 Optical scheme to observe dispersion of light by a prism.

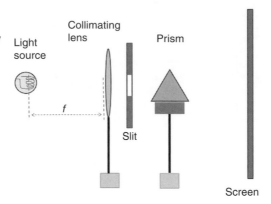

Figure 16.5 Incandescent light source connected to the power supply.

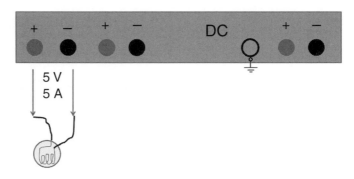

(b) Light Emitting Diode as Light Source

1. Replace the incandescent bulb with a light emitting diode, and design the electrical circuit shown in Fig. 16.7.

2. Adjust the distance between the light emitting diode and the collimating lens (Fig. 16.4) to produce a collimated beam. Some diodes are designed to produce quasi-collimated light. If this is the case, you do not need a collimator.

3. Observe the light passing through the prism and landing on the screen. Briefly describe your observations.

(c) Laser as Light Source

1. Replace the incandescent bulb with a red laser. Expand the laser beam using a laser beam expander, as shown in Fig. 16.8.

   Lens 1: Focal length is ~12–18 mm;

Dispersion of Light and Prism Spectroscope

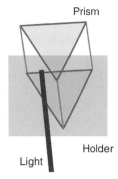

Figure 16.6 Relative position of the prism and incident light.

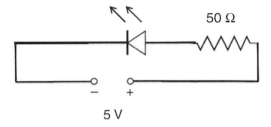

Figure 16.7 Light emitting diode.

Lens 2: Focal length is ~100 mm.

2. Adjust the distance between the lenses (Fig. 16.8) to produce collimated light.

3. Observe the light passing through the prism onto the screen. Mark the position of the laser light on the screen using a needle. Briefly describe your observations.

4. Replace the red laser with a green laser and repeat steps 2 and 3.

5. Draw a conclusion about both the dispersion of light by a prism and the spectral composition of the light sources.

How Color and Neutral Absorbing Filters Work

1. Assemble the optical set-up shown in Fig. 16.4 (use an incandescent light source).

2. Insert a color filter between the slit and the prism, as shown in Fig. 16.9.

Figure 16.8 Optical scheme to observe dispersion and/or deviation of light by a prism.

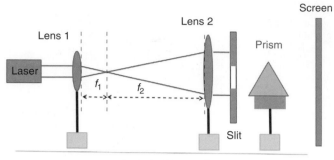

Figure 16.9 Optical set-up to observe changes in light spectrum caused by absorption.

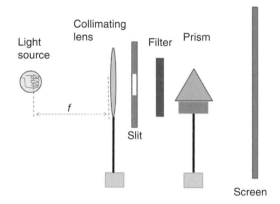

3. Observe changes in the light spectrum on the screen.

4. Replace the color filter with a neutral filter and again observe changes in the light spectrum on the screen.

5. Explain the observed phenomenon and draw a conclusion about how absorbing filters work.

Observation of Normal Dispersion of Light Using Two Crossed Prisms

1. Assemble the optical set-up shown in Fig. 16.10 (use an incandescent light source).

2. Use two identical equal-angle prisms (#1 and #2, as shown in Fig. 16.10). It is important that prism 1 is "perpendicular" to prism 2. Prism 1 splits falling "white" light into rainbow colors,

Dispersion of Light and Prism Spectroscope

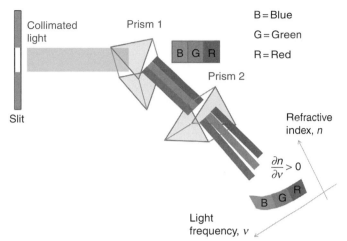

Figure 16.10 Method of crossed prisms to observe normal dispersion of light (for simplicity, only red, green, and blue colors are shown, while intermediate colors are omitted).

which are deviated by prism 2 depending on the dispersion of the prism material.

3. Explain the shape of the light which emerges from prism 2.

4. Discuss the limitations of this method. Propose a way to make quantitative measurements of the refractive index using the "method of crossed prisms."

Designing a Prism Spectroscope

A prism spectroscope is composed of a light source, slit, collimator, prism (dispersive element), telescope, and an angle-measuring scale with verniers. The collimator, prism, and telescope are mounted on a solid table so that they can be rotated around the axis (the so-called mechanical instrument axis) perpendicular to the table surface. A light source is used to illuminate the slit, and the collimator collimates light that emerges from the slit and falls onto the prism. The prism is a dispersive element. It splits falling light into its spectrum. The telescope analyzes the light emerging from the prism and can select certain spectral lines which can be observed with the eye. Let us describe some basic components of the spectroscope as separate procedures.

(a) Illumination of the Slit

Proper illumination of the spectroscope slit is achieved as described below:

1. If the light source is bright enough, place the light source in close contact with and in front of the slit, as shown in Fig. 16.11.

Figure 16.11 Relative positions of the light source, slit, and collimating lens.

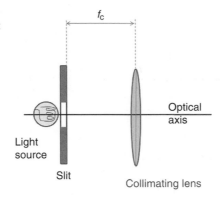

Figure 16.12 Illumination of a slit using an auxiliary lens.

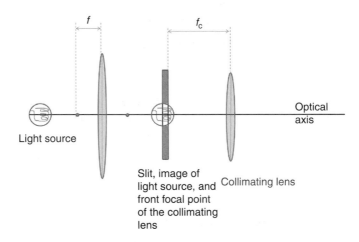

The collimating lens, or collimator, should be placed in such a way that the slit position coincides with the front focal point of the collimator.

2. If the light source is not bright enough, an auxiliary lens is used to collect light emitted by the light source. This lens forms an image of the light source either magnified or minified, depending on the sizes of the bulb filament and slit. If the bulb filament is smaller than the slit height, the image is magnified. The position of the image coincides with the position of the slit, as shown in Fig. 16.12. The collimating lens, or collimator, should be placed in such a way that the slit position coincides with the front focal point of the collimator.

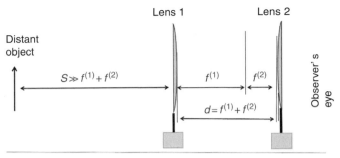

Figure 16.13 Optical telescope.

3. Identify appropriate lenses in your optics kit and assemble the optical set-up shown in Fig. 16.11, using an incandescent bulb as the light source. Analyze the light emerging from the collimating lens.

4. Assemble the optical set-up shown in Fig. 16.12. Analyze the light emerging from the collimating lens.

5. Draw a conclusion about different ways to illuminate the slit.

(b) Focusing Telescope at Infinity

1. Identify appropriate lenses (Lens 1: focal length ~50–100 mm; lens 2: focal length ~12–18 mm) and assemble an optical set-up as shown in Fig. 16.13 to make an optical telescope.

2. Focus the telescope at infinity. To do this, choose any distant object and bring its image into focus by adjusting the distance between lenses 1 and 2.

(c) Arrangement of the Collimator, Prism, and Telescope

1. On a solid table, place the collimator and the telescope focused for parallel light, as described in previous procedures. Place the prism on the prism table.

2. Assemble the optical set-up shown in Fig. 16.14.

3. The optical axes of the collimator and the telescope should lie in the same plane. Additionally, these optical axes must be perpendicular to the mechanical instrumental axis. The prism table can be rotated around this instrumental axis.

4. Observe the image of the slit using the telescope. Focus the telescope for red light, then rotate the telescope and bring into focus the image of the slit for green light. Finally, bring into focus a blue image of the slit.

Figure 16.14 Principal optical scheme of the prism spectroscope.

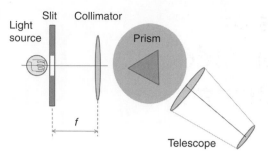

Evaluation and Review Questions

1. Derive the expression for the chromatic resolving power of the prism.
2. Develop an experimental procedure to graduate a prism spectroscope to measure the wavelength of light.

Further Reading

General

E. Hecht, *Optics*, 4th edition, San Francisco, CA: Addison-Wesley, 2001

F. L. Pedrotti, S. J. L. Pedrotti, L. M. Pedrotti, *Introduction to Optics*, 3rd edition, Upper Saddle River, NJ: Pearson Prentice Hall, 2007

J. Strong, *Concepts of Classical Optics*, New York: Dover, 2004 (Dover edition is an unabridged republication of the work originally published in 1958 by W. H. Freeman and Company, San Francisco, CA)

Specialized

Handbook of Optics, W. G. Driscoll (editor), W. Vaughan (associate editor), New York: McGraw-Hill, 1978

R. Ditteon, *Modern Geometrical Optics*, New York: John Wiley & Sons, 1998

D. C. O'Shea, *Elements of Modern Optical Design*, New York: Wiley, 1985

C. H. Palmer, *Optics: Experiments and Demonstrations*, Baltimore, MD: Johns Hopkins Press, 1962

A. F. Wagner, *Experimental Optics*, New York: John Wiley & Sons, 1929

Elements of Computer-Aided Optical Design

17

Objectives

To develop basic skills to handle and use popular software packages such as:

- ASAP
- Code V
- OSLO
- FRED
- TracePro
- Zemax
- KDP-2
- OpTaliX

Background

Consider an optical system composed of lenses. An ideal optical system is expected to form a perfect image when the image of an object passes through the optical system with uniform magnification. Separate optical elements (lenses, mirrors) fail to build such a perfect optical image due to the optical aberrations in these elements. Since optical aberrations can have both positive and negative signs, a proper combination of lenses can eliminate certain types of optical aberrations. In addition, lens bending (choosing appropriate surface shapes) and insertion of diaphragms (stops) in suitable positions are two widely used methods of reducing aberrations and improving image quality. *Optical design* deals with the variety of methods used to make a perfect real optical system. A "perfect real optical system" is: (a) An optical system consisting of a minimum, but necessary, number of optical elements which are correctly bent. In other words, the elements have special surface shapes and thicknesses and are made of the correct optical materials; are held in the right positions and with diaphragms inserted where needed; (b) An optical system that performs close to that of the ideal optical system (light point is imaged into light point).

Methods of optical design are based on both geometrical and physical optics. In the following example, we focus on "geometrical" optical design. One reason to use a "geometrical" description of an optical system is that image blurring caused by optical aberrations exceeds image blurring caused by the diffraction of light.

Any optical system has optical components (lenses, mirrors, etc.) and mechanical components (lens mounting, holders, etc.). Optical components are made of special *optical materials*. Choosing the right optical material depends on the goals of an optical system. When choosing optical materials, the following details should be taken into consideration: (1) Light wavelength (e.g. UV, visible, IR) and transmission; (2) light flux, low power or high power (this factor determines the optical threshold of the material); (3) dispersive properties; (4) optical quality of the material (is there staining, weathering, bubbles?); (5) thermal properties (refractive index depends on temperature); (6) availability; (7) price. Optical companies have their own *catalogs of optical materials* (both printed and electronic) used to make optical elements. Optical designers should be familiar with these catalogs. In addition, catalogs of optical materials are incorporated into *modern optical software*, or special programs developed to design optical systems.

Once optical materials are chosen, the performance of the optical system should be checked (it is assumed that we already know the principal scheme of the optical system). First, *paraxial ray tracing* is checked. Since a real optical system operates with non-paraxial light rays, the second activity is to make an *exact ray tracing*. Currently, exact ray tracing (as well as paraxial ray tracing) is done by using *special programs developed for optical design*. The modern market in *computer-aided optical design* provides both commercial and free programs and is highly competitive and evolves quickly. Because it is difficult to provide a complete list of the currently available optical programs, we will limit ourselves to the most popular software packages: ASAP, Code V, OSLO, FRED, TracePro, Zemax, KDP-2, and OpTaliX. A variety of similar programs and their detailed descriptions can be found using web resources.

Now we will describe the general principles of computer-aided optical design. The systematic method of lens design in use today is an iterative technique. Computer-aided optical design includes the following steps:

1. Choosing the lens type (radius, thickness, diameter), lens material, and the total number of lenses.

2. Paraxial ray tracing.

3. Exact ray tracing and calculation of the *merit function*.

4. Minimization of the merit function using appropriate mathematical methods (damped least squares or orthonormalization of aberrations) and creation of the improved optical system.

5. Optimization of the improved optical system.

6. Tolerance analysis.

Exact ray tracing allows for the determination of the Seidel aberration coefficients (see Chapter 12 for details) which measure image quality. Instead of using words such as "good" or "bad" to describe the image, and, therefore, "good" or "bad" lens, optical designers introduced the already-mentioned *merit function*. This is a special single number used to represent lens quality. Most merit functions are really "demerit functions," which represent the sum of the squares of various image errors: Merit function $= \mathrm{MF} = \sum_{i=1}^{n} \omega_i d_i^2$, where ω is the statistical "weight" of the defect item d. The larger the merit function value, the worse the image.

The following are examples of the merit function. Because Seidel aberration coefficients (S, C, A, P, D) are widely used, the merit function in this case can be written as $\mathrm{MF} = S^2 + C^2 + A^2 + P^2 + D^2$. This expression treats all aberrations equally. For cases when some aberrations are more important than others, an improved merit function can be created:

$$\mathrm{MF} = \omega_1 (S - S_0)^2 + \omega_2 (C - C_0)^2 + \omega_3 (A - A_0)^2 + \omega_4 (P - P_0)^2 + \omega_5 (D - D_0)^2 + \omega_6 (F - F_0)^2$$

where S_0, C_0, A_0, P_0, D_0, and F_0 are the target values for the five Seidel coefficients and the f-number of the system, and ω_1 through ω_6 are the weights assigned to each of the terms in the merit function.

The merit function can be modified depending on the particular case. Because of this, an optical designer should always check what merit function is required when using an optical design program.

The minimization of the merit function can be done in the following manner. First, the input optical system is ray traced, and the merit function is computed. Then, one of the parameters which describe the optical system is changed, and the merit function is calculated again. After changing the parameter series, a table is created to express merit function changes versus parameter changes. By using some mathematical technique (for example, damped least squares or orthonormalization of aberrations), an improved optical system is created. The resulting improved optical system should have a minimum value equal to the merit

Figure 17.1
Illustration of tilt (Δθ)
and decentration
(ΔY): (a) Optical
system is perfectly
aligned; (b) plano-
convex lens is tilted,
and plano-concave
lens is decentrated.

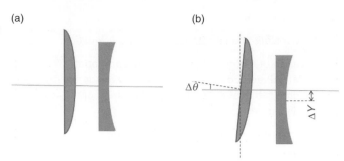

function. The total computer time needed to complete these calculations is proportional to the number of rays and input parameters. Improved optical systems can be optimized by removing or adding additional elements. Note that only experienced optical designers know where and when to add or remove optical elements. A computer only helps make routine and technical calculations. A bad input will certainly produce a bad output.

So far, we have discussed the optical part of the optical system. Now, let us switch our attention to the mechanical part of the optical system. This is also important because all optical components are held by means of certain mechanical holders. Perfectly designed optical systems should remain image stable and without visible changes when small *tilts* Δθ and *decentrations* ΔY take place (Fig. 17.1). Such stability of the optical system can be evaluated with a so-called *tolerance analysis*.

Figure 17.1 also illustrates the importance of the mounting and centering of the optical system. In fact, mounting and centering are separate parts of optical design. Most optical design programs include the ability to specify tilts and decentrations as part of a tolerance analysis.

The ultimate goal of an optical design is to achieve an image of good quality. Therefore, any optical design program deals with *image evaluation*.

The performance of an optical system can be evaluated using a so-called *spot diagram*: A rectangular grid is imposed on the entrance pupil of the system. Next, ray tracing is completed for rays coming from an object point and passing through the intersections of the grid lines in the pupil. Then, the locations of these rays on a plane in image space (as a rule, at the paraxial image plane) are plotted. For an ideal optical system, all the rays would intersect at one point: the image point. In reality, because of the optical aberrations, the rays are spread out in a region in the vicinity of the image point. This spreading represents a *spot diagram* which provides insight into the performance of an optical system. Figure 17.2 shows an example of a spot diagram calculated using an optical design program.

(a)

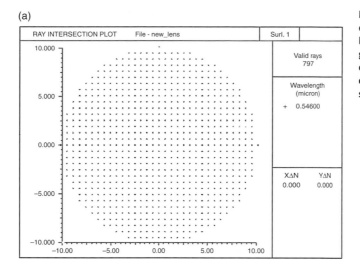

(b)

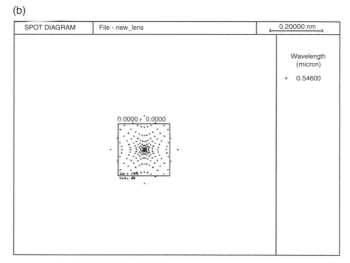

Figure 17.2 Spot diagram for a single lens: (a) Rectangular grid imposed on the entrance pupil of the optical system; (b) spot diagram vs. field.

In addition to spot diagrams, more solid image evaluations can be made using the *point spread function* (PSF) and the *modulation transfer function* (MTF).

The PSF describes the response of an imaging system to a point source or point object.

The MTF takes into account the extended nature of objects. It is a measure of the accuracy with which different frequency components are reproduced in the image. The MTF evaluates the quality of the image of a multiple-bar object.

A final note is that optical design is an art based on fundamental science.

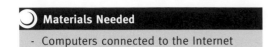

Materials Needed

- Computers connected to the Internet

Procedures

Analysis of the Current Optical Software Market

Use the web links given at the end of this chapter to analyze the current status of the optical software market. Make classifications of the available optical design programs. Notice advantages and disadvantages of the available optical design programs. Complete the following table.

Table 17.1

Optical design program	Vendor and web link	Program's features	Availability and price (if specified)	Comments

Fundamentals of OpTaliX

- Download and install a trial version of the optical design program OpTaliX. In addition, download the program manual and tutorial for this program.

- Read the manual and tutorial and develop a procedure to analyze an optical system using this program.

- As a simple example, consider two cases: (a) A single lens and (b) a lens doublet.

- For these two cases, make a complete analysis of the optical elements. Present the results of your studies as a procedure.

Note: Your analysis should include lens parameters (radii of curvature, thickness, lens material), spot diagrams, exact ray tracing, lens layout, Seidel coefficients, PSF, and MTF.

Fundamentals of KDP-2

Repeat the same procedure for the KDP-2 program.

Fundamentals of Zemax

Repeat the same procedure for the Zemax program.

Evaluation and Review Questions

Think about how to handle aspheric surfaces using optical design software.

Conclusion

Compare the programs studied in this chapter and formulate their advantages and disadvantages.

Further Reading

Specialized

R. Ditteon, *Modern Geometrical Optics*, New York: John Wiley & Sons, 1998

R. Kingslake, R. B. Johnson, *Lens Design*, New York: Academic Press, 2010

M. Laikin, *Lens Design*, New York: Marcel Dekker, 2001

D. C. O'Shea, *Elements of Modern Optical Design*, New York: Wiley, 1985

W. J. Smith, *Modern Optical Engineering*, 3rd edition, New York: McGraw Hill, 2000

Web Links

www.breault.com/software/asap.php

www.ecalculations.com/

www.lambdares.com

www.lambdares.com/oslo (OSLO package)

www.occuvu.com.au/opticalCalculations.php (optical calculator)

www.optenso.com/download/download.html (OpTaliX)

www.optenso.com/links/links.html (links on optical design and engineering)

www.opticalres.com/cv/cvprodds_f.html

www.zemax.com/

Index

Index